12th Edoardo Amaldi Conference on Gravitational Waves (AMALDI 12)

AA006743

Journal of Physics: Conference Series Volume 957

Pasadena, California, USA
9 - 14 July 2017

Editor:

Sydney Meshkov

ISBN: 978-1-5108-5930-2
ISSN: 1742-6588

Printed from e-media with permission by:

Curran Associates, Inc.
57 Morehouse Lane
Red Hook, NY 12571

Some format issues inherent in the e-media version may also appear in this print version.

Copyright© (2017) by the Institute of Physics
All rights reserved. The material featured in this book is subject to
IOP copyright protection, unless otherwise indicated.

Printed by Curran Associates, Inc. (2018)

For permission requests, please contact the Institute of Physics
at the address below.

Institute of Physics
Dirac House, Temple Back
Bristol BS1 6BE UK

Phone: 44 1 17 929 7481
Fax: 44 1 17 920 0979

techtracking@iop.org

Additional copies of this publication are available from:

Curran Associates, Inc.
57 Morehouse Lane
Red Hook, NY 12571 USA
Phone: 845-758-0400
Fax: 845-758-2633
Email: curran@proceedings.com
Web: www.proceedings.com

Table of contents

Volume 957

12th Edoardo Amaldi Conference on Gravitational Waves (AMALDI 12)
9–14 July 2017, Pasadena, CA, USA

Accepted papers received: 16 January 2018
Published online: 5 March 2018

Preface

Papers

IOP Conf. Series: Journal of Physics: Conf. Series **957** (2018) 011001 doi:10.1088/1742-6596/957/1/011001

The Edoardo Amaldi Conference on Gravitational Waves is the premier forum for the fields of gravitational wave science and gravitational wave detection. Held biannually, the Amaldi meetings are organized under the auspices of the Gravitational Wave International Committee (GWIC), which oversees the scientific program for the meeting.

Amaldi 12 was held at the Hilton Hotel in Pasadena, CA from July 9-14, 2017. This was the first Amaldi conference since LIGO and Virgo announced the detection of gravitational waves from the merger of two black holes into a single black hole. Unlike previous Amaldi Conferences that focused on anticipatedgravitational wave science and detector development, this was the first Amaldi conference to examine the science arising from the LIGO-Virgo detections and the successful LISA Pathfinder mission, the precursor to LISA.

During the conference, the full range of gravitational wave science was represented, including plans for the further development of terrestrial and space based detectors, as well as progress in pulsar timing arrays. Emphasis was placed on the connection between gravitational wave science and the fields of observational astrophysics and multi-messenger astronomy, as well as in progress in modeling the waveforms for gravitational waves arising from various astrophysical sources, highlighting the stronger connection between experimental advances, numerical relativity simulations, and astrophysics theory.

Guest editor
Dr. Sydney Meshkov
Caltech, LIGO 100 – 36, Pasadena, CA 91125 (syd@ligo.caltech.edu)

Content from this work may be used under the terms of the Creative Commons Attribution 3.0 licence. Any further distribution of this work must maintain attribution to the author(s) and the title of the work, journal citation and DOI.
Published under licence by IOP Publishing Ltd

IOP Conf. Series: Journal of Physics: Conf. Series **957** (2018) 011002 doi:10.1088/1742-6596/957/1/011002

Peer review statement

All papers published in this volume of *Journal of Physics: Conference Series* have been peer reviewed through processes administered by the proceedings Editors. Reviews were conducted by expert referees to the professional and scientific standards expected of a proceedings journal published by IOP Publishing.

Content from this work may be used under the terms of the Creative Commons Attribution 3.0 licence. Any further distribution of this work must maintain attribution to the author(s) and the title of the work, journal citation and DOI.
Published under licence by IOP Publishing Ltd

Black hole demography at the dawn of gravitational-wave astronomy: state-of-the art and future perspectives

Michela Mapelli[1,2]

[1] INAF – Osservatorio Astronomico di Padova, Vicolo dell'Osservatorio 5, I–35122, Padova, Italy
[2] Institute for Astrophysics and Particle Physics, University of Innsbruck, Technikerstrasse 25/8, A–6020 Innsbruck, Austria

E-mail: `michela.mapelli@oapd.inaf.it`

Abstract. The first four LIGO detections have confirmed the existence of massive black holes (BHs), with mass $30-40$ $M_{\odot}$. Such BHs might originate from massive metal-poor stars ($Z < 0.3$ $Z_{\odot}$) or from gravitational instabilities in the early Universe. The formation channels of merging BHs are still poorly constrained. The measure of mass, spin and redshift distribution of merging BHs will give us fundamental clues to distinguish between different models. In parallel, a better understanding of several astrophysical processes (e.g. common envelope, core-collapse SNe, and dynamical evolution of BHs) is decisive, to shed light on the formation channels of merging BHs.

1. Lesson learned from the first four detections

Four detections of black hole (BH) mergers have been reported so far by the LIGO-Virgo collaboration and several tens of new detections are expected in the next few years [1, 2, 3, 4, 5, 6]: the era of gravitational wave astronomy has just begun.

The first four detections already have crucial implications for the astrophysics of BHs: they confirm the existence of BH binaries and tell us that such systems can host massive BHs, i.e. BHs with mass higher than ~ 20 $M_{\odot}$.

This was a big surprise for the astrophysics community, considering that the only BHs for which we had a dynamical mass measurement before LIGO's detection were about a dozen of BHs in nearby X-ray binaries and their mass is well below 20 $M_{\odot}$ [7, 8].

Moreover, the first four detections also provide us with a measurement of the mass and spin of the final (i.e. post-merger) BH and with some (still rather limited) information on the spins of the merging BHs. Finally, the estimate of the redshift of merging BHs represents the first step towards reconstructing the cosmic BH merger rate. Table 1 is a summary of the main quantities obtained from LIGO-Virgo observations with their 90% credible intervals (according to [3, 5, 6]). In the next sections, we will review our state-of-the-art knowledge of BH astrophysics and we will discuss the implications of the first four detections on the mass spectrum of BHs and on the formation channels of mergig BHs.

Content from this work may be used under the terms of the Creative Commons Attribution 3.0 licence. Any further distribution of this work must maintain attribution to the author(s) and the title of the work, journal citation and DOI.
Published under licence by IOP Publishing Ltd

IOP Conf. Series: Journal of Physics: Conf. Series **957** (2018) 012001 doi:10.1088/1742-6596/957/1/012001

2. Implications for the mass spectrum of black holes

Three of the first four detected events (GW150914, GW170104, and GW170814) contain BHs with mass $m_{\rm BH} > 30$ M$_\odot$. If these BHs have stellar origin, such large masses can be understood only in terms of quenched stellar winds and failed supernova (SN) explosions.

2.1. Supernovae

Our understanding of core-collapse SNe is still far from satisfactory. The outcome of a core-collapse SN depends on the mechanism of neutrino transfer, as well as on the impact of stellar rotation and magnetic fields [9]. Moreover, two- and three-dimensional hydrodynamical simulations are still too computationally expensive to be used for a systematic study of the remnant mass (e.g. [10]). Despite these issues, recent models of core-collapse SNe have started converging on some common ground.

The models of [11] suggest that the final mass of the BH depends mainly on the mass of the Carbon-Oxygen (CO) core and on the amount of fallback. For CO core masses $m_{\rm CO} \gtrsim 8$ M$_\odot$ a star is expected to directly collapse to a BH, without SN explosion. In this case, the entire final mass of the star collapses to a BH, producing a more massive BH than in case of a SN explosion.

Several authors suggest that the CO mass is not sufficient to understand the final fate of a star. For example, [12, 13] introduce the compactness parameter, defined as

$$\xi_M = \frac{M/{\rm M}_\odot}{R(M)/1000\,{\rm km}}, \qquad (1)$$

where M is a reference mass (usually $M = 2.5$ M$_\odot$) and $R(M)$ is the radius which encloses this mass. The compactness parameter is usually measured at the onset of collapse [13]. A large value of ξ_M ($\xi_{2.5} \gtrsim 0.2$, accoding to [14]) indicates that the star is going to collapse to a BH directly.

Limongi (2017, [15]) has shown that there is a strong correlation between the final CO mass and the value of $\xi_{2.5}$ (see figure 20 of [15]), suggesting that models based on the CO mass and on the compactness parameter are in reasonable agreement. The main difference is that models based on the compactness parameter [12, 13] and even more complex models (e.g. the two parameter criterion by [10]) predict islands of "explodability" (i.e. even very massive stars might have $\xi_{2.5} < 0.2$ and thus are expected to explode as SNe), while models based on the CO mass [11] predict a rather smooth transition between exploding and non-exploding models (as a function of the final mass of the star).

These differences must not be neglected, but overall we can predict a general trend:

- The details of the SN mechanism and the efficiency of fallback are particularly important for stars with mass $<< 30$ M$_\odot$, which are not particularly affected by stellar winds.
- For massive (> 30 M$_\odot$) stars, stellar winds are extremely important and determine the final fate of a star.

In addition, pulsational-pair instability and pair instability SNe play a crucial role for very massive metal poor stars [16]:

- Stars with $30 \lesssim M_{\rm He}/{\rm M}_\odot \lesssim 60$ (where $M_{\rm He}$ is the mass of the Helium core), are expected to undergo pulsational pair instability, losing a large fraction of their mass.
- If $60 \lesssim M_{\rm He}/{\rm M}_\odot \lesssim 130$, a star is expected to explode as pair-instability SN, leaving no remnant.
- For $M_{\rm He} \gtrsim 130$ M$_\odot$, the star is so massive that the collapse induced by pair production cannot be reversed: the star is expected to directly collapse to a very massive BH.

Figure 1 clearly shows the impact of pair instability and pulsational pair instability SNe on the final remnant mass.

Table 1. Main observed properties of current GW detections.

	GW150914	LVT151012	GW151226	GW170104	GW170814
m_1 (M$_\odot$)	$36.2^{+5.2}_{-3.8}$	23^{+18}_{-6}	$14.2^{+8.3}_{-3.7}$	$31.2^{+8.4}_{-6.0}$	$30.5^{+5.7}_{-3.0}$
m_2 (M$_\odot$)	$29.1^{+3.7}_{-4.4}$	13^{+4}_{-5}	$7.5^{+2.3}_{-2.3}$	$19.4^{+5.3}_{-5.9}$	$25.3^{+2.8}_{-4.2}$
$m_{\rm chirp}$ (M$_\odot$)	$28.1^{+1.8}_{-1.5}$	$15.1^{+1.4}_{-1.1}$	$8.9^{+0.3}_{-0.3}$	$21.1^{+2.4}_{-2.7}$	$24.1^{+1.4}_{-1.1}$
$m_{\rm TOT}$ (M$_\odot$)	$65.3^{+4.1}_{-3.4}$	37^{+13}_{-4}	$21.8^{+5.9}_{-1.7}$	$50.7^{+5.9}_{-5.0}$	$55.9^{+3.4}_{-2.7}$
$m_{\rm fin}$ (M$_\odot$)	$62.3^{+3.7}_{-3.1}$	35^{+14}_{-4}	$20.8^{+6.1}_{-1.7}$	$48.7^{+5.7}_{-4.6}$	$53.2^{+3.2}_{-2.5}$
$\chi_{\rm eff}$	$-0.06^{+0.14}_{-0.14}$	$0.0^{+0.3}_{-0.2}$	$0.21^{+0.20}_{-0.10}$	$-0.12^{+0.21}_{-0.30}$	$0.06^{+0.12}_{-0.12}$
$S_{\rm f}$	$0.68^{+0.05}_{-0.06}$	$0.66^{+0.09}_{-0.10}$	$0.74^{+0.06}_{-0.06}$	$0.64^{+0.09}_{-0.20}$	$0.70^{+0.07}_{-0.05}$
$d_{\rm L}$ (Mpc)	420^{+150}_{-180}	1000^{+500}_{-500}	440^{+180}_{-190}	880^{+450}_{-390}	540^{+130}_{-210}
z	$0.09^{+0.03}_{-0.04}$	$0.20^{+0.09}_{-0.09}$	$0.09^{+0.03}_{-0.04}$	$0.18^{+0.08}_{-0.07}$	$0.11^{+0.03}_{-0.04}$

m_1 (m_2): mass of the primary (secondary) BH; $m_{\rm chirp}$ ($m_{\rm TOT}$): chirp (total) mass of the binary; $m_{\rm fin}$: mass of the final BH; $\chi_{\rm eff}$: effective spin; $S_{\rm fin}$: spin of the final BH; $d_{\rm L}$: luminosity distance; z: redshift. From left to right: GW150914, LVT151012, GW151226, GW170104, GW170814.

2.2. Stellar winds

Mass loss by stellar winds is important because it determines the final mass of a star: according to the core-collapse SN models we discussed in the previous section, stars with a larger final mass ($\gtrsim 30$ M$_\odot$, [17, 18]) are more likely to undergo a failed SN, leaving a larger remnant.

In the last decade, it has become increasingly clear that stellar winds depend on metallicity, not only during the main sequence (MS) but also afterwards [19, 20, 21, 22]. Another important player is the Eddington factor $\Gamma = L_*/L_{\rm Edd}$ (where L_* and $L_{\rm Edd}$ are the luminosity of the star and its Eddington limit, respectively). Most population-synthesis models account for the metallicity dependence of stellar winds, but neglect the impact of Γ, with few remarkable exceptions (e.g. [23, 24, 25]). According to Chen et al. (2015, [26]), mass loss by stellar winds can be modelled as $\dot{M} \propto Z^\alpha$, with

$$\alpha = \begin{cases} 0.85 & \text{if}\,\Gamma \leq 2/3 \\ 2.45 - 2.4\,\Gamma & \text{if}\,\Gamma > 2/3 \end{cases} \tag{2}$$

This indicates that the dependence of $\dot{M}$ on metallicity becomes negligible if a star is radiation pressure dominated (i.e. Γ approaches 1).

Figure 1 shows the mass spectrum of BHs obtained with the mass-loss prescription given in equation 2 plus the delayed core-collapse SN model explained in [11]. According to this model, the lower the metallicity is, the larger the maximum possible mass of the compact remnant.

Recent studies [27] show that a surface magnetic field has a similar effects to metallicity, quenching mass loss. Also, rotation cannot be neglected: on the one hand, stellar winds are enhanced in fast rotating massive stars, on the other hand chemical mixing induces the formation of a larger Helium (and Carbon-Oxygen) core [15]. While stronger stellar winds should translate into smaller remnant masses (see previous subsetion), larger core masses might imply larger remnant masses. To complicate the situation even more, pair instability and pulsational pair instability SNe are expected to occur even at smaller zero age MS (ZAMS) mass if the Helium core mass is larger, possibly reducing the number and mass of BHs [15].

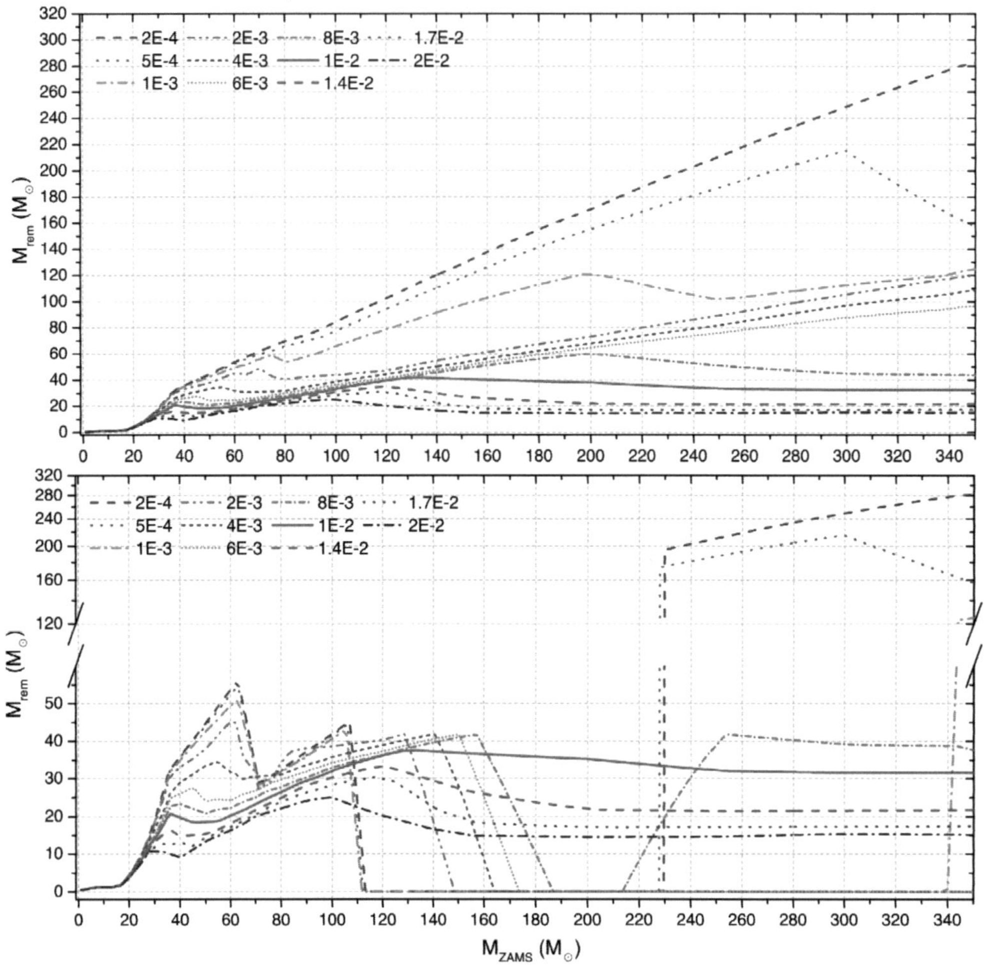

Figure 1. Mass of the compact object as a function of the ZAMS mass of the progenitor star for 11 different metallicities. Upper (lower) panel: pair instability and pulsational pair instability SNe are not (are) included in the model. Adapted from Figures 1 and 2 of [24].

2.3. Primordial BHs

Primordial BHs (e.g. [28, 29, 30]) are predicted to form from gravitational instabilities in the very early Universe (< 1 sec before the Big Bang) and their allowed mass range is $\sim 1 - 1000$ $M_\odot$ [29][1]. Thus, they might have a mass consistent with the observed gravitational wave events.

2.4. Take home message

In summary, our knowledge of the BH mass spectrum is still hampered by large uncertainties (on stellar winds, on SNe, and on the existence of primordial BHs). However, we can conclude that the progenitors of GW150914, GW170104 and GW1708114 were relatively metal-poor massive stars (e.g. [4, 31, 32, 24]), unless we invoke primordial BHs or BHs born from previous mergers [33].

[1] Considering only BHs which produce gravitational waves in the LIGO-Virgo frequency band when they merge

3. Field or dynamical origin?

The mechanisms driving the formation of close BH binaries are still matter of intense debate. Two main channels have been proposed: the evolution of isolated BH binaries and the dynamical formation in star clusters.

3.1. Isolated binary evolution

According to the isolated binary evolution, two massive stars form gravitationally bound and then they evolve through several binary evolution processes, till they become BHs and (if they are sufficiently close) they merge. Unfortunately, we have a rather poor knowledge of several binary evolution processes, especially mass transfer and common envelope [34]. Two stars with a ZAMS mass $\gtrsim 30$ M$_\odot$ are expected to reach photospheric radii of several thousand solar radii when they become super-giant stars. If the initial semi-major axis of a binary is of the same order of magnitude as the maximum stellar radii, the binary is expected to prematurely merge, or to transfer mass through Roche-lobe overflow (if one of the two stars fills its Roche lobe), or to undergo a common envelope phase (e.g. if both stars fill their Roche lobes). During common envelope, the cores of the two stars are engulfed in the same envelope and are expected to spiral in, because of gas drag. If nothing removes the envelope, the two cores keep spiralling in till they merge: the binary produces only a single BH. In contrast, if the common envelope is ejected (e.g. by conversion of the orbital energy of the cores into thermal energy), the final binary is a naked binary (without envelope) and has a very short final semi-major axis. If SN kicks do not disrupt it, the naked binary might evolve into a double BH binary, sufficiently close to merge in a Hubble time. In contrast, binaries with much larger initial semi-major axes remain 'detached binaries', i.e. they avoid mass transfer and common envelope, but are too loose to produce merging BHs.

Models used in population-synthesis codes do not capture the complex physics of common envelope. The most used formalism depends on two free parameters (α and λ, describing the efficiency of conversion of orbital energy into thermal energy and the binding energy of the envelope, respectively [35]), which are uncertain to a factor of $\gtrsim 10$.

Alternative models [36, 37, 38] predict that massive fast rotating metal-poor stars might produce merging BH binaries even without common envelope evolution. In fact, fast rotating stars might remain almost chemically homogeneus during their evolution. As a consequence, their stellar radii remain significantly smaller than in the non-rotating case. The chemically homogeneous evolution model can account only for the formation of nearly equal-mass merging BH binaries, with large mass (BH masses > 20 M$_\odot$) and aligned spins.

In contrast, the common-envelope model (e.g. [39, 31, 40, 41, 25]) predicts the formation of merging BH binaries with both small and large masses (BH masses > 3 M$_\odot$), accounting for both GW150914-like systems and GW151226-like systems. Strongly inequal-mass systems ($m_2/m_1 < 0.5$) are quite unlikely even in the common-envelope model. Common-envelope and tides are expected to align the spins of the two stars with the orbital angular momentum of the binary, although SN explosions or other mechanisms might disalign the final BH spins (e.g. [42, 43]).

3.2. Dynamical formation

According to the dynamical formation scenario, two BHs can enter the same binary through a dynamical mechanism (for example, a dynamical exchange [44]). Dynamical mechanisms (with the exception of Kozai-Lidov resonance, [45, 46]) are efficient only if stars are in dense environments, such as young star clusters, open clusters, globular clusters and nuclear star clusters. On the other hand, massive stars (which are BH progenitors) form preferentially in dense environments [47], suggesting that dynamical mechanisms are crucial for the formation and evolution of BH binaries.

Three-body encounters (i.e. close encounters between a binary system and an intruder star) affect the orbital properties of binaries, such as semi-major axis and eccentricity. Hard binaries (i.e. binaries with binding energy larger than the average kinetic energy of intruders) are expected to *harden* as a consequence of three-body encounters, meaning that their semi-major axis shrinks [48]. As a consequence of dynamical hardening, a BH binary might become sufficiently close as to merge by gravitational wave emission in a Hubble time. Thus, dynamics can boost the merger rate.

Exchanges are three-body encounters during which the intruder replaces one of the former members of the binary. The probability of an exchange is particularly high if the mass of the intruder is larger than the mass of one of the members of the binary [44]. Thus, BHs and BH progenitors, being among the most massive stars in star clusters, are particularly effective in capturing companions through exchanges. Ziosi et al. (2014, [49]) find that $\gtrsim 90$ per cent of double BH binaries in young star clusters and open clusters form through dynamical exchanges.

Three-body encounters and exchanges have a crucial impact on a BH binary population: they tend to produce more massive binaries, with non-zero eccentricity, and to randomize their spins [49, 50, 51, 32, 52, 53, 54]. A nearly isotropic distribution of BH spins is expected from dynamics.

In young massive metal-poor clusters, dynamics can even lead to the formation of intermediate-mass BHs (i.e. BHs with mass $10^2 \leq m_{\rm BH}/{\rm M}_\odot \leq 10^4$), through the *runaway collision mechanism* [55, 56, 32].

Unlike the previous processes, the Kozai-Lidov (KL) resonance [45, 46] might be important also in low-density environments. A hierarchical triple undergoes KL resonance if the inclination between the orbital plane of the inner binary and that of the outer body is non zero. Perturbations by the outer body induce periodic oscillations of both the eccentricity of the inner binary and the inclination between the two orbital planes, without changing the semi-major axis. The increase of eccentricity might speed up the merger of the inner binary and might even produce systems with non-zero eccentricity in the LIGO-Virgo frequency range [57, 58, 59]. KL resonances are deemed to be a quite common process, since the occurrence of triple systems in massive stars might be as high as ~ 50 per cent (see [60] and references therein). In addition, [57] propose that KL oscillations might also occur in very exotic triple systems, composed of an inner BH binary and (as a tertiary body) a super-massive BH, in nuclear star clusters.

In summary, dynamics has multiple effects on merging BHs: (i) dynamics tends to produce more massive BH binaries, or even intermediate-mass BH binaries; (ii) dynamically formed binaries might have non zero eccentricity, even close to the merger; (iii) we expect a nearly isotropic distribution of spins in dynamically formed binaries; (iv) dynamics should boost the merger rate (e.g. through hardening).

Unfortunately, our knowledge of the dynamical formation scenario is hampered by several issues. While many papers have been published about the dynamics of BHs in old globular clusters (e.g. [61, 62, 63, 64, 65, 50, 51, 66, 67]) only few papers deal with young star clusters (e.g. [49, 32, 52, 53]), which are the most common birthplace of massive stars, at least in the nearby Universe. The main reason is likely a numerical one: globular clusters are spherically symmetric, relaxed, gasless old systems which can be simulated with a fast Monte Carlo approach, while young star clusters are rather asymmetric, substructured, not yet relaxed, gas rich systems, requiring accurate and expensive direct N-body simulations to study them. Moreover, there is no complete study of young star cluster dynamics in a cosmological context. For these reasons, constraints on the dynamical formation scenario are still quite poor.

3.3. Merger rates from dynamically formed versus isolated binaries

The BH merger rate inferred from current LIGO-Virgo detections is $\sim 12 - 213$ Gpc^{-3} yr^{-1} [5]. Several studies report merger rates from their models, which can be compared to this estimate.

IOP Conf. Series: Journal of Physics: Conf. Series **957** (2018) 012001 doi:10.1088/1742-6596/957/1/012001

- *Isolated binaries.* Local merger rates from isolated binaries span from $R_{\rm BHB} \sim 10$ to ~ 800 Gpc^{-3} yr^{-1} [39, 31, 41], depending on the assumptions about common-envelope efficiency, SN kicks and distribution of BH progenitors with redshift (see e.g. Figure 2).
- *Globular clusters.* Merger rates from BH binaries in globular clusters are estimated to be $R_{\rm BHB} \sim 5$ [51, 67]. This value is still uncertain but appears to be quite low with respect to the inferred rate. Actually, we expect the merger rate to be boosted in globular clusters by dynamical effects, but globular clusters represent a small fraction of the total stellar mass in galaxies (less than few per cent, [68]).
- *Young star clusters.* Merger rates from young star clusters and open clusters are extremely uncertain, because there are only few hundreds N-body simulations of these systems. The allowed range currently spans $\sim 0.1 - 100$ Gpc^{-3} yr^{-1} [49, 52, 69].
- *Nuclear star clusters.* The rate of mergers happening in nuclear star clusters was estimated to be 1-2 Gpc^{-3} yr^{-1} by [70]. However, this number is also subject to large uncertainties. Most mergers in nuclear star clusters seem to involve relatively massive systems (similar to GW150914).

3.4. Take-home message
From this Section, we have seen that masses and spins of merging BHs might help us disentangling between the isolated binary evolution scenario and the dynamical formation scenario (e.g. from dynamics we expect more massive binaries with uniform spin distribution), but still too many issues and uncertainties affect theoretical models, especially the dynamical evolution. New more sofisticated models are needed, especially for young star clusters. It is also important to apply model selection (see e.g. [66]) between all existing models, in order to understand how many detections are needed to reject a given model.

4. The cosmological context
The first four observed events have redshift spanning from ~ 0.09 to ~ 0.2. With the increase of sensitivity in the next observing runs, LIGO and Virgo will be able to observe BH mergers even at higher redshifts, while the third generation ground-based gravitational wave detectors (e.g. Einstein Telescope, [71]) will enable us to reconstruct the entire merger history of BH binaries up to redshift $\sim 8 - 10$.

A lot of work has been done recently to investigate the evolution of the BH merger rate with redshift (e.g. [72, 31, 73, 74, 41]). Several studies suggest that the BH merger rate in the comoving frame is going to increase with redshift (Fig. 2) with a trend depending on both the cosmic star-formation rate density and the metallicity evolution [72, 41].

This result is possibly very important for constraining BH formation channels. In fact, while the merger rate of stellar-born BHs approximately follows the cosmic star formation rate, there is no reason for primordial BHs to behave in the same way (although the details depend on the delay time between the formation of the BHs and their merger). The key question here is how many detections we need to reconstruct the cosmic BH merger rate "sufficiently well" and what is the minimum sensitivity interferometers should reach to probe a large enough volume. Is design sensitivity of Advanced LIGO and Virgo enough to get some constraints from the merger history or should we wait for third generation ground-based detectors?

5. Conclusions
The first four LIGO detections have shown that stellar BHs might be quite more massive than we thought ($\sim 30 - 40$ M$_\odot$). Models suggest that such massive BHs can arise only from the evolution of relatively metal-poor massive stars ($Z \lesssim 0.3$ Z$_\odot$ [18, 75, 76, 11, 77, 23]), or from gravitational instabilities in the early Universe (primordial BHs, [29]).

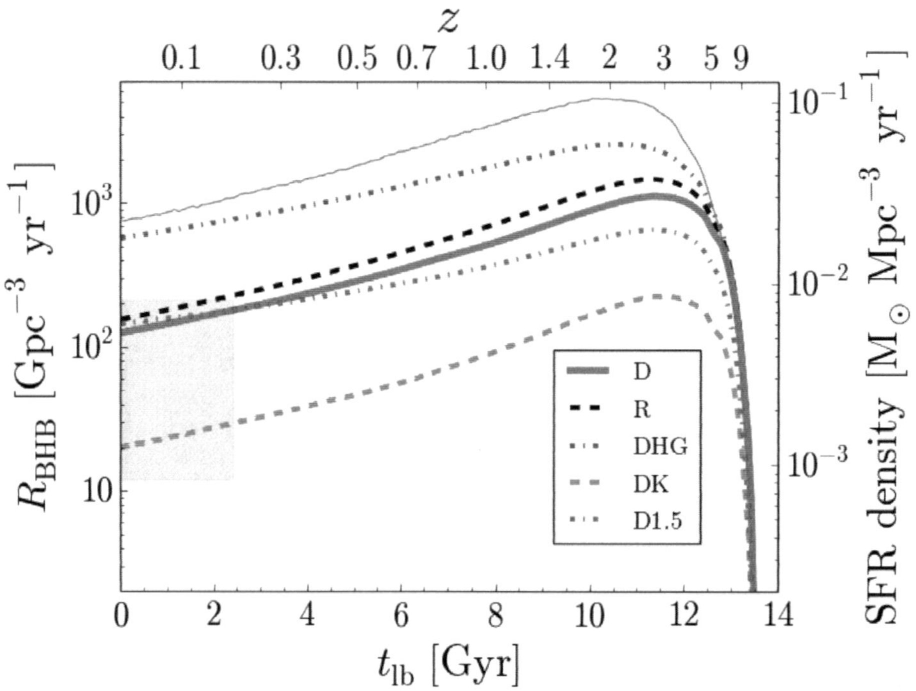

Figure 2. Cosmic merger rate density of BH binaries in the comoving frame ($R_{\rm BHB}$) as a function of the look-back time ($t_{\rm lb}$, bottom x−axis) and of the redshift (z, top x−axis) for several population synthesis models [25] interfaced with the Illustris simulation. Red solid line: model D (delayed SN model [11], CE parameters $\alpha\,\lambda = 0.1$, fiducial distribution for the natal kicks of the BHs). Black dashed line: model R, same as model D but for a rapid SN model [11]. Green dot-dashed line: model D1.5 (same as model D but for $\alpha\,\lambda = 1.5$). Violet dot-dashed line: model DHG (same as model D but assuming that Hertzsprung gap donors can survive a CE phase). Orange dashed line: model DK (same as model D but with maximum BH natal kicks). Right-hand y−axis: cosmic star formation rate density in the Illustris, shown as a thin grey line. Adapted from Figure 1 of [41].

The cosmic merger rate of stellar-born BHs is expected to approximately scale with the cosmic star formation rate density [72, 41], while the merger rate of primordial BHs is expected to follow a different trend. Future detections might be able to disentangle between stellar-born BHs and primordial BHs, if the cosmic BH merger rate history is reconstructed to a sufficiently high redshift.

Not only BH masses but also BH spins can give us possible clues to understand the formation channels of merging BHs (dynamical evolution model versus isolated binary evolution). However, the uncertainty on measured effective spins are still very large and models are also affected by several problems. For example, it is not clear how the magnitude of the spin of a BH depends on the magnitude of the spin of the progenitor star [33]. While it is known that dynamically formed binaries prefer misaligned spins, there is still no astrophysically-motivated distribution for the spins in isolated binaries. These uncertainties result in several degeneracies between models.

The era of gravitational-wave astrophycs has now begun, opening a completely new perspective for the study of BH demography. To meet this challenge, theoretical models must reach a better understanding of the astrophysical processes driving the formation of BH binaries, both for the dynamical and for the isolated formation channels.

Acknowledgments
I thank the organizers and the participants of the 12th Edoardo Amaldi Conference on Gravitational Waves for the enlightening discussions. Numerical calculations have been performed through a CINECA-INFN agreement and through a CINECA-INAF agreement (Accordo Quadro INAF-CINECA 2017), providing access to resources on GALILEO and MARCONI at CINECA. MM acknowledges financial support from the Italian Ministry of Education, University and Research (MIUR) through grant FIRB 2012 RBFR12PM1F, from INAF through grant PRIN-2014-14, and from the MERAC Foundation.

References
[1] Abbott, B. P., Abbott, R., Abbott, T. D., et al. 2016, Physical Review Letters, 116, 061102
[2] Abbott, B. P., Abbott, R., Abbott, T. D., et al. 2016, Physical Review Letters, 116, 241103
[3] Abbott, B. P., Abbott, R., Abbott, T. D., et al. 2016, Physical Review X, 6, 041015
[4] Abbott, B. P., Abbott, R., Abbott, T. D., et al. 2016, Astrophysical JournalLetter, 818, L22
[5] Abbott, B. P., Abbott, R., Abbott, T. D., et al. 2017, Physical Review Letters, 118, 221101
[6] Abbott, B. P., Abbott, R., Abbott, T. D., et al. 2017, Physical Review Letters, 119, 141101
[7] Özel, F., Psaltis, D., Narayan, R., & McClintock, J. E. 2010, Astrophysical Journal, 725, 1918
[8] Farr, W. M., Sravan, N., Cantrell, A., et al. 2011, Astrophysical Journal, 741, 103
[9] Foglizzo, T., Kazeroni, R., Guilet, J., et al. 2015, Publications of the Astronomical Society of Australia, 32, e009
[10] Ertl, T., Janka, H.-T., Woosley, S. E., Sukhbold, T., & Ugliano, M. 2016, Astrophysical Journal, 818, 124
[11] Fryer, C. L., Belczynski, K., Wiktorowicz, G., et al. 2012, Astrophysical Journal, 749, 91
[12] O'Connor, E., & Ott, C. D. 2011, Astrophysical Journal, 730, 70
[13] Ugliano, M., Janka, H.-T., Marek, A., & Arcones, A. 2012, Astrophysical Journal, 757, 69
[14] Horiuchi, S., Nakamura, K., Takiwaki, T., Kotake, K., & Tanaka, M. 2014, MNRAS, 445, L99
[15] Limongi, M. 2017, arXiv:1706.01913
[16] Woosley, S. E. 2017, Astrophysical Journal, 836, 244
[17] Fryer, C. L. 1999, Astrophysical Journal, 522, 413
[18] Mapelli, M., Colpi, M., & Zampieri, L. 2009, MNRAS, 395, L71
[19] Vink, J. S., de Koter, A., & Lamers, H. J. G. L. M. 2001, Astronomy & Astrophysics, 369, 574
[20] Vink, J. S., & de Koter, A. 2005, Astronomy & Astrophysics, 442, 587
[21] Gräfener, G., & Hamann, W.-R. 2008, Astronomy & Astrophysics, 482, 945
[22] Vink, J. S., Muijres, L. E., Anthonisse, B., et al. 2011, Astronomy & Astrophysics, 531, A132
[23] Spera, M., Mapelli, M., & Bressan, A. 2015, MNRAS, 451, 4086
[24] Spera, M., & Mapelli, M. 2017, MNRAS, 470, 4739
[25] Giacobbo, N., Mapelli, M., & Spera, M. 2018, MNRAS, 474, 2959
[26] Chen, Y., Bressan, A., Girardi, L., et al. 2015, MNRAS, 452, 1068
[27] Petit, V., Keszthelyi, Z., MacInnis, R., et al. 2017, MNRAS, 466, 1052
[28] Bird, S., Cholis, I., Muñoz, J. B., et al. 2016, Physical Review Letters, 116, 201301
[29] Carr, B., Kühnel, F., & Sandstad, M. 2016, Physical Review Documents, 94, 083504
[30] Inomata, K., Kawasaki, M., Mukaida, K., Tada, Y., & Yanagida, T. T. 2017, Physical Review Documents, 95, 123510
[31] Belczynski, K., Holz, D. E., Bulik, T., & O'Shaughnessy, R. 2016, Nature, 534, 512
[32] Mapelli, M. 2016, MNRAS, 459, 3432
[33] Gerosa, D., & Berti, E. 2017, Physical Review Documents, 95, 124046
[34] Ivanova, N., Justham, S., Chen, X., et al. 2013, The Astronomy and Astrophysics Review, 21, 59
[35] Webbink, R. F. 1984, Astrophysical Journal, 277, 355
[36] Marchant, P., Langer, N., Podsiadlowski, P., Tauris, T. M., & Moriya, T. J. 2016, Astronomy & Astrophysics, 588, A50
[37] de Mink, S. E., & Mandel, I. 2016, MNRAS, 460, 3545
[38] Mandel, I., & de Mink, S. E. 2016, MNRAS, 458, 2634
[39] Dominik, M., Belczynski, K., Fryer, C., et al. 2012, Astrophysical Journal, 759, 52

[40] Stevenson, S., Vigna-Gómez, A., Mandel, I., et al. 2017, Nature Communications, 8, 14906
[41] Mapelli, M., Giacobbo, N., Ripamonti, E., & Spera, M. 2017, MNRAS, 472, 2422
[42] O'Shaughnessy, R., Gerosa, D., & Wysocki, D. 2017, Physical Review Letters, 119, 011101
[43] Wysocki, D., Gerosa, D., O'Shaughnessy, R., et al. 2017, arXiv:1709.01943
[44] Hills, J. G., & Fullerton, L. W. 1980, Astronomical Journal, 85, 1281
[45] Kozai, Y. 1962, Astronomical Journal, 67, 591
[46] Lidov, M. L. 1962, Planetary and Space Science, 9, 719
[47] Portegies Zwart, S. F., McMillan, S. L. W., & Gieles, M. 2010, Annual Review of Astronomy and Astrophysics, 48, 431
[48] Heggie, D. C. 1975, MNRAS, 173, 729
[49] Ziosi, B. M., Mapelli, M., Branchesi, M., & Tormen, G. 2014, MNRAS, 441, 3703
[50] Rodriguez, C. L., Morscher, M., Pattabiraman, B., et al. 2015, Physical Review Letters, 115, 051101
[51] Rodriguez, C. L., Chatterjee, S., & Rasio, F. A. 2016, Physical Review Documents, 93, 084029
[52] Banerjee, S. 2017, MNRAS, 467, 524
[53] Banerjee, S. 2018, MNRAS, 473, 909
[54] Chatterjee, S., Rodriguez, C. L., & Rasio, F. A. 2017, Astrophysical Journal, 834, 68
[55] Portegies Zwart, S. F., Baumgardt, H., Hut, P., Makino, J., & McMillan, S. L. W. 2004, Nature, 428, 724
[56] Giersz, M., Leigh, N., Hypki, A., Lützgendorf, N., & Askar, A. 2015, MNRAS, 454, 3150
[57] Antonini, F., & Perets, H. B. 2012, Astrophysical Journal, 757, 27
[58] Kimpson, T. O., Spera, M., Mapelli, M., & Ziosi, B. M. 2016, MNRAS, 463, 2443
[59] Antonini, F., Toonen, S., & Hamers, A. S. 2017, Astrophysical Journal, 841, 77
[60] Toonen, S., Hamers, A., & Portegies Zwart, S. 2016, Computational Astrophysics and Cosmology, 3, 6
[61] Sigurdsson, S., & Hernquist, L. 1993, Nature, 364, 423
[62] Portegies Zwart, S. F., & McMillan, S. L. W. 2000, Astrophysical JournalLetter, 528, L17
[63] Downing, J. M. B., Benacquista, M. J., Giersz, M., & Spurzem, R. 2010, MNRAS, 407, 1946
[64] Downing, J. M. B., Benacquista, M. J., Giersz, M., & Spurzem, R. 2011, MNRAS, 416, 133
[65] Tanikawa, A. 2013, MNRAS, 435, 1358
[66] Zevin, M., Pankow, C., Rodriguez, C. L., et al. 2017, Astrophysical Journal, 846, 82
[67] Askar, A., Szkudlarek, M., Gondek-Rosińska, D., Giersz, M., & Bulik, T. 2017, MNRAS, 464, L36
[68] Harris, W. E., Harris, G. L. H., & Alessi, M. 2013, Astrophysical Journal, 772, 82
[69] Fujii, M. S., Tanikawa, A., & Makino, J. 2017, Publications of the Astronomical Society of Japan, 69, 94
[70] Antonini, F., & Rasio, F. A. 2016, Astrophysical Journal, 831, 187
[71] Punturo, M., Abernathy, M., Acernese, F., et al. 2010, Classical and Quantum Gravity, 27, 194002
[72] Dominik, M., Belczynski, K., Fryer, C., et al. 2013, Astrophysical Journal, 779, 72
[73] Lamberts, A., Garrison-Kimmel, S., Clausen, D. R., & Hopkins, P. F. 2016, MNRAS, 463, L31
[74] Schneider, R., Graziani, L., Marassi, S., et al. 2017, MNRAS, 471, L105
[75] Mapelli, M., Ripamonti, E., Zampieri, L., Colpi, M., & Bressan, A. 2010, MNRAS, 408, 234
[76] Belczynski, K., Bulik, T., Fryer, C. L., et al. 2010, Astrophysical Journal, 714, 1217
[77] Mapelli, M., Zampieri, L., Ripamonti, E., & Bressan, A. 2013, MNRAS, 429, 2298

IOP Conf. Series: Journal of Physics: Conf. Series **957** (2018) 012002 doi:10.1088/1742-6596/957/1/012002

The payloads of Advanced Virgo: current status and upgrades

L. Naticchioni[1,2] ***on behalf of the Virgo Collaboration***
[1]Università di Roma *Sapienza*, I-00185 Rome, Italy
[2]INFN - Sezione di Roma, I-00185 Rome, Italy
E-mail: `luca.naticchioni@roma1.infn.it`

Abstract. The development and integration of new detector payloads has been an important part of the Advanced Virgo (AdV) project, the major upgrade of the Virgo interferometric detector of Gravitational Waves, aiming to increase the detector sensitivity by one order of magnitude. During the integration phase of the new AdV payloads with monolithic suspension of mirrors we experienced systematic suspension failures later identified as caused by dust contamination of the vacuum system. In order to not postpone the detector commissioning, making possible to join the LIGO O2 observation run, the Collaboration decided to proceed with the integration of the payloads relying on steel wire suspensions for all the mirrors. In this proceeding the status of the currently integrated payloads is reported, including their angular control characterization and the Q-factor measurements for test mass steel wire suspensions. The payload upgrade for the re-integration of monolithic suspensions after the O2 run is reported in the last section.

1. Introduction

The first observations of gravitational waves (GW) [1, 2, 3] has been possible using large-scale Michelson interferometers (ITF) in which the variation of the light path due to passing GWs is detected as a fluctuation in the dark fringe signal. The largest ITF GW detectors built so far are the two 4 km-long LIGO observatories in USA and the 3 km-long European observatory Virgo in Italy. In these detectors the mirrors are suspended from seismic insulators in order to act as gravitational test masses. During the last years both LIGO and Virgo have been updated to their Advanced design. The Advanced Virgo [4] project (AdV) started in 2011 and was completed this year. The aim of this upgrade is the improvement of the detector sensitivity by one order of magnitude, reached using heavier mirrors and higher circulating laser power among other relevant changes. These requirements led to redesign the last suspension stage for the mirrors of AdV, referred to as payload. It is a double pendulum stage consisting of marionette, mirror (i.e. the test mass, TM), the associated suspensions and actuators. The payload is in turn suspended from the Superattenuator (SA) in a dedicated UHV chamber located at the base of the SA tower [4]. It is conceived for the compensation of the residual seismic noise and for the steering and control of the main optical components of the ITF.

In initial Virgo and Virgo+ payloads [5] the mirror was suspended from the marionette in parallel with a recoil mass, used for its actuation, thus forming a double stage branched pendulum [7]. In the current Advanced Virgo design this configuration was changed into a simple double stage suspension, where the mirror actuation is obtained using coils fixed to the

Content from this work may be used under the terms of the Creative Commons Attribution 3.0 licence. Any further distribution of this work must maintain attribution to the author(s) and the title of the work, journal citation and DOI.
Published under licence by IOP Publishing Ltd

same mechanical structure that is also used for the marionette actuation, the so called Actuator Cage. This structure is fixed to the last steering filter of the SA and enfolds the payload, providing the support frame for baffles, thermal compensation elements (compensation plate and ring heater) [4] and acting also as safety frame for the suspended elements of the payload. In order to reduce the suspension thermal noise, the input and end mirrors of the ITF Fabry-Perot cavities are monolithically suspended from the marionette using four fused silica fibers for each mirror [8]. The removal of the recoil mass in Advanced Virgo payloads has the advantage of simplifying the payload structure and its local control, moreover it implies the reduction of thermal noise associated with the recoil mass suspension. On the other hand the simple double stage design reduces also the vertical and horizontal degrees of inertial damping provided by the suspension, that is however still compliant with the residual seismic noise requirements at the level of the mirror [6]. A direct comparison between the payloads of Virgo/Virgo+ and Advanced Virgo is shown in fig. 1. A technical drawing of the AdV Payload integrated with the SA suspension chain is shown in fig. 2.

Figure 1. Comparison between Virgo+ and Advanced Virgo payloads. Main elements: 1) interface with the last SA steering filter, 2) Marionette, 3) Actuator Cage, 4) Recoil Mass, 5) Mirror.

Figure 2. CAD drawings of the Advanced Virgo payload integrated in the SuperAttenuator suspension.

2. AdV Payloads in O2

Before the completion of the AdV project, seven new payloads have been integrated in Virgo: the beam splitter payload (BS), two input payloads (NI, North Input and WI, West Input), two end payloads (NE, North End and WE, West End), the Power Recycling (PR) and the Signal Recycling (SR) payloads. Even though they all share the same concept design, each type of payload is characterized by different details, as described in the next subsections. In all the payloads the marionette, suspended with a single maraging steel wire, has an octagonal shape and is made of 316L amagnetic steel plus high density copper-tungsten alloy counterweights. Moreover two counterweights can be moved along orthogonal axis by remotely controlled stepper motors, for a fine tuning of the horizontal balancing of the marionette. Input and end payloads are assembled adopting the monolithic suspension of mirrors with SiO_2 fibers [8], BS and recycling mirrors are suspended with steel wires instead. Both the pendulum stages are contained inside the Actuator Cage (fig. 3). At its top this structure has a disk-shaped element that supports the coils of the marionette actuators (coil disk in fig. 3). At both sides facing the mirror the cage is designed to support large diaphragm baffles made of stainless steel plates coated with Diamond-like Carbon that are used to reduce stray light in the main beams. The

IOP Conf. Series: Journal of Physics: Conf. Series **957** (2018) 012002 doi:10.1088/1742-6596/957/1/012002

lower part of the cage, the so called cradle, acts as a safety-stop structure for the mirror. During the integration with the SA suspension the coil disk is fixed to an interface structure that in turn is connected to the last SA steering filter (fig. 2), from which the marionette is then suspended with a steel wire. During the first vacuum operations before ITF commissioning the monolithic suspensions of input and end payloads were affected by failures, whose investigations and conclusions are given in [9]. While the tests were underway, the Collaboration decided to re-assemble input and end mirrors in their relative payloads using steel wires, in order to complete the ITF commissioning in time to join LIGO during the O2 observation run.

2.1. Beamsplitter Payload
The BS payload has been designed to host the largest mirror ever suspended in an ITF GW detector, having a diameter of 550 mm. This optical element is suspended from the marionette by four steel wires of section 300 μm. The BS payload weighs 100.5 kg and was the first to be integrated during the AdV upgrade in December 2014.

2.2. Input Payloads
The input payloads are designed for the suspension of the new AdV input test masses (ITM), that have the same diameter of initial Virgo (i.e. 350 mm), but a doubled thickness (200 mm), thus a weight of 42 kg. Considering also the marionette, the input payload weighs 145.2 kg. Their Actuator Cage (AC) is designed to host the frames and positioning mechanisms for two important elements of the thermal compensation system (TCS [4], fig. 3,4): the Compensation Plate (CP) and the Ring Heater (RH). Their fine positioning is remotely controlled by stepper motors. As in the other payload cages, the AC acts also as safety structure and support frame for marionette and mirror actuators and baffles. The AC coil disk includes a balancing mass at the opposite side of the compensation plate. Payload assembly and integration phases are shown in fig. 4.

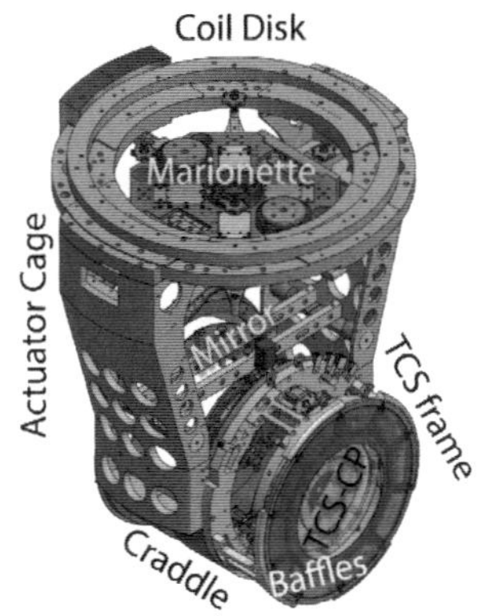

Figure 3. Input Payload, CAD drawing.

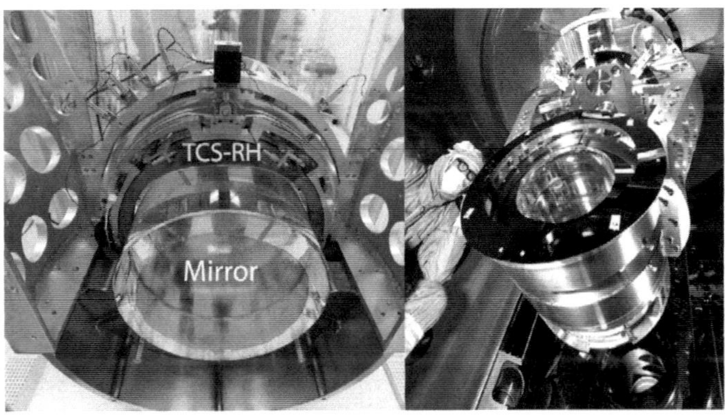

Figure 4. Input Payload during assembly (left) and its integration with SA (right).

2.3. End Payload
End payloads are similar to the input ones since the end test masses (ETM) have the same dimensions and weight (42 kg) of ITMs. Conversely their AC differs from that of input payloads, since it does not host the compensation plate support frame. However the ring heater frame and its positioning mechanism are present also in their AC.

IOP Conf. Series: Journal of Physics: Conf. Series **957** (2018) 012002 doi:10.1088/1742-6596/957/1/012002

2.4. Recycling Payloads
The power recycling (PR) and signal recycling (SR) payloads are conceptually similar to input payloads but in this case the mirrors have half the thickness and the same diameter of ITM and ETM, for a weight of 21 kg, resulting in a payload overall weight of 85.2 kg. The AC is similar to the others, providing a support frame for marionette and mirror actuators, baffles and safety stops. The cage hosts also a tilted frame for an additional optical element, the pick-off plate, used to extract the pick-off signal from the PR mirror [4]. Also the inclination of the plane of the pick-off plate with respect to the mirror can be changed by means of two stepper motors.

3. ITM and ETM payloads characterization
The angular position of the suspended payload elements (marionette and mirror) is controlled by the suspension Local Control. Given the reference frame reported in fig. 2, the error signals are produced by a set of optical levers (OpL) measuring pitch (θ_x), yaw (θ_y) and roll (θ_z) of every marionette and pitch, yaw and pendulum mode (z) of every mirror. In particular in this section we report the angular characterization of ITM and ETM payloads and the preliminary measurement of the quality factor of the current steel wire suspension.

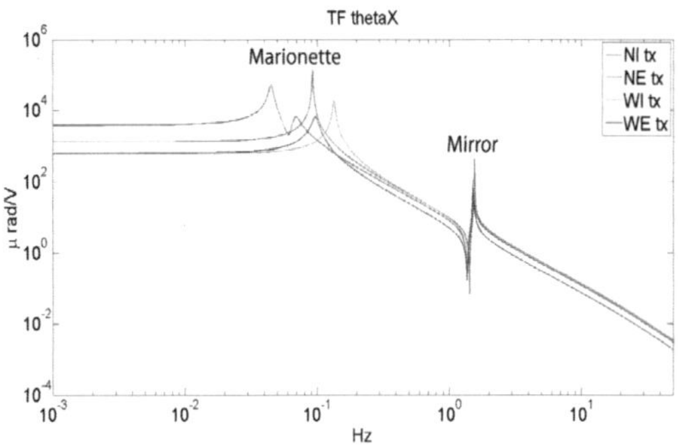

Figure 5. Pitch transfer functions for ITM (NI,WI) and ETM (NE,WE) payloads measured between marionette and mirror.

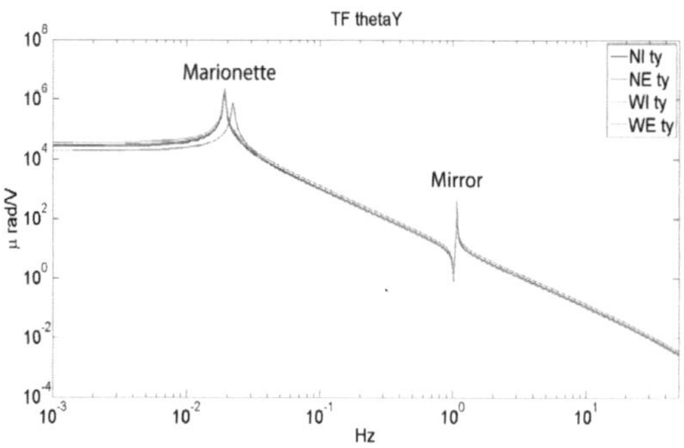

Figure 6. Yaw transfer functions for ITM (NI,WI) and ETM (NE,WE) payloads measured between marionette and mirror.

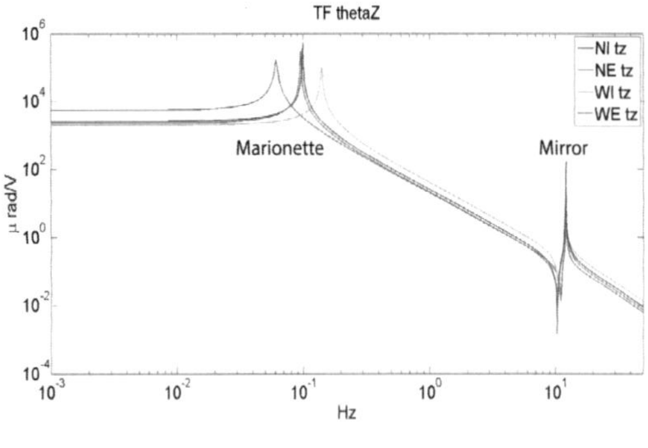

Figure 7. Roll transfer functions for ITM (NI,WI) and ETM (NE,WE) payloads measured between marionette and mirror.

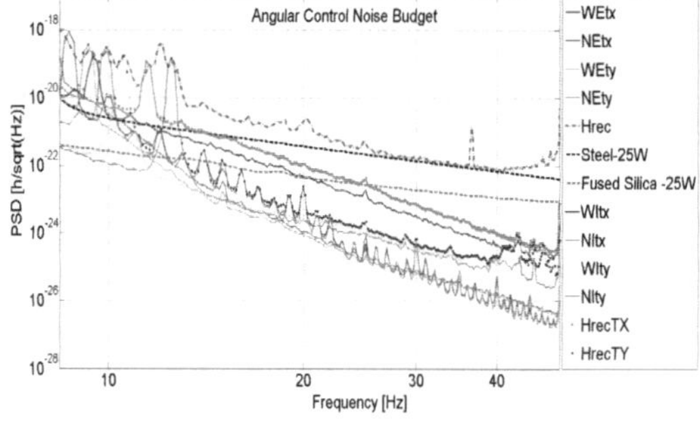

Figure 8. Angular control noise budget of TM payloads below 50 Hz, compared to dark fringe signal and target sensitivities.

3.1. Angular Transfer Functions

The angular transfer functions (TF) between marionette and mirror of each ITM and ETM payload were measured relying on the angular error signals given by the OpLs. Pitch, yaw and roll are reported respectively in fig. 5, 6 and 7. In the frequency band 300 - 800mHz the microseismic noise is transferred to the payload by the notches present in the SA transfer function and then it may impact on payload lowest modes of both pitch and roll. Therefore the payloads were designed to have those modes well below that frequency range. As choosing the backup solution of steel wire mirror suspension for the O2 run, we explored the possibility to further lowering those modes in the case of NI payload (fig. 5 and 7), successfully attaining smooth and clean operation. This feature will be adopted also during the re-integration of the monolithic suspensions.

3.2. Angular noise

Given the transfer functions between the local control and the strain (h reconstructed, i.e. the signal coming out of the ITF), the correction signal to be applied during standard operation can be easily projected, obtaining the angular noise budget reported in fig. 8. Here it is compared with the ITF target sensitivities, given a laser power of 25 W, in two cases: steel-wire suspension and monolithic TM suspension. The dark fringe signal (Hrec) achieved during the O2 run is also reported in the same figure, proving to be not limited by the angular control noises (HrecTX, HrecTY), while the spectral lines around 10 Hz are injected and demodulated to allow the ITF beam centering on the mirrors. The angular control is based upon both quadrant photodiodes (QPHD) and OpLs: a further improvement of QPHD sensors and digital filters of the control loops is foreseen after the O2 run, reducing the angular noise to allow the ITF target sensitivity.

3.3. Steel wires Q-factor

The quality factor of the steel-wire suspension of ITM and ETM was estimated injecting kicks with the motorized counterweight in the payload marionette and measuring the amplitude exponential decay of the four spectral peaks produced by the first violin mode of the wires of each payload. Indeed, the decay time τ is obtained from the exponential fit of each peak decay, and then the Q-factor is given by $Q = -\pi f \tau$. Preliminary results are given in table 1 and indicate Q-factor values ranging from 2 to 3 $\times 10^5$, in accordance with expectations for steel wires. Frequency and Q-factor ranges refer to the four wires of the single payload.

Table 1. Preliminary results of Q-factors and loss angles of the steel wire suspension of ITM and ETM in Advanced Virgo measured before O2 run.

Payload	Kick GPS time	Frequency range	Q-factor range	Loss angle ϕ
NE	1179043885	$306.6 - 314.5 Hz$	$(2.21 - 2.37) \times 10^5$	$\sim 1.8 \times 10^{-3}$
NI	1179045541	$306.7 - 313.6 Hz$	$(2.18 - 2.71) \times 10^5$	$\sim 1.7 \times 10^{-3}$
WI	1179047015	$307.3 - 311.8 Hz$	$(2.03 - 2.69) \times 10^5$	$\sim 1.5 \times 10^{-3}$
WE	1179048159	$308.8 - 311.7 Hz$	$(1.94 - 2.92) \times 10^5$	$\sim 2.0 \times 10^{-3}$

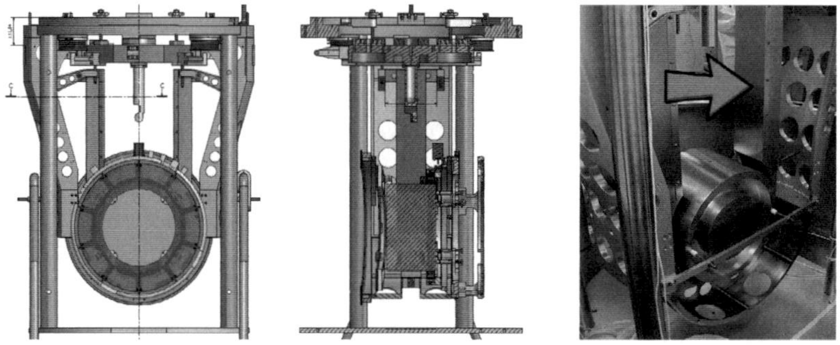

Figure 9. Test payload with monolithic suspension and FG integrated, indicated by red and purple in the CAD drawing and by the red arrow in the picture.

4. ITM and ETM payloads upgrade after O2

Before the O3 observation run, ITM and ETM payloads will be dismounted and their mirror steel-wire suspension will be replaced by the monolithic suspension. The complete procedure of payload extraction, disassembly, re-assembly with monolithic suspension and reintegration in the SA will require about three weeks for each payload. In addition to the planned improvements of vacuum system and cleaning procedures, a new component called Fiber Guard (FG, fig. 9) will be added to the payload AC to avoid any possible fused silica fiber failure due to the impact of high speed particles related to vacuum operations. FGs are made of thin Al-alloy and are designed to shield up to $\sim$ 90% of the fibers surface, to be compatible with the local control FOV and to not change significantly the overall suspended weight. FGs will be integrated in the AC immediately after the installation of fused silica fibers in the payload. Prototype FGs have been already integrated in a monolithic test payload (fig. 9). FEM simulation were produced to estimate the frequency shift of the AC resonant modes due to the introduction of FGs, obtaining a negligible change: less than 3% for modes below $250Hz$.

5. Conclusions

Advanced Virgo payloads integration started in 2014 and was concluded this year. TM payloads were re-integrated with steel wire suspensions after the monolithic suspension failures due to vacuum system contaminations [9]. Their angular transfer functions were measured as well as the angular control noise projection. As expected for steel wires, the Q-factor of ITM and ETM payloads is in the range of 2×10^5. During the stop after the O2 observation run, TM payloads will be dismounted and re-assembled with the monolithic suspension. To prevent any possible further failure due to high speed particles during vacuum operations, among other improvements of the vacuum system, fused silica fibers will be shielded by the fiber guards, a new component of the payload actuator cage. Their integration with a monolithic suspension has been already done in a test payload.

References

[1] Abbott BP et al. 2016 *Phys. Rev. Lett.* **116** 061102
[2] Abbott BP et al. 2017 *Phys. Rev. Lett.* **119** 141101
[3] Abbott BP et al. 2017 *Phys. Rev. Lett.* **119** 161101
[4] Acernese F et al. 2014 *Class. Quantum Grav.* B **32** 024001
[5] Bernardini A, et al. 1999 *Rev. Sci. Instrum.* **70** 3463
[6] Accadia T et al. 2011 *J. Low Freq. Noise V. A.* **30** 63
[7] Accadia T et al. 2012 *J. Instrum.* **7** P03012
[8] Aisa D et al. 2016 *Nucl. Instrum. Methods Phys. Res. A* **824** 644
[9] Travasso F et al. 2017, in preparation, to be published in this volume of *J. Phys. Conf. Ser.*

Prospects for gravitational wave astronomy with next generation large-scale pulsar timing arrays

Yan Wang[1] and Soumya D. Mohanty[2]

[1] MOE Key Laboratory of Fundamental Physical Quantities Measurements, Hubei Key Laboratory of Gravitation and Quantum Physics, School of Physics, Huazhong University of Science and Technology, 1037 Luoyu Road, Wuhan, Hubei Province 430074, China

[2] Department of Physics and Astronomy, The University of Texas Rio Grande Valley, One West University Blvd, Brownsville, TX 78520, USA

E-mail: ywang12@hust.edu.cn; soumya.mohanty@utrgv.edu

Abstract.
Next generation radio telescopes, namely the Five-hundred-meter Aperture Spherical Telescope (FAST) and the Square Kilometer Array (SKA), will revolutionize the pulsar timing arrays (PTAs) based gravitational wave (GW) searches. We review some of the characteristics of FAST and SKA, and the resulting PTAs, that are pertinent to the detection of gravitational wave signals from individual supermassive black hole binaries.

1. Introduction

The high-frequency ($10 - 10^4$ Hz) gravitational wave (GW) window in observational astronomy and cosmology has been opened by LIGO with the first direct detection of GWs from the merger of a stellar-mass black hole binary [1]. Progress on opening up the GW window at lower frequencies continues [2; 3] with the $10^{-9} - 10^{-6}$ Hz being probed by pulsar timing arrays (PTAs). This frequency regime is expected to host GW signals from supermassive black hole binaries residing in the innermost regions of merged galaxies. Other potential signals that could be detected are stochastic GW background from cosmic strings [4], inflation [5] and first-order QCD phase transition [6].

Three main regional collaborations are involved in advancing the search for GWs with PTAs: the North American Nanohertz Observatory for Gravitational Waves (NANOGrav) [7] based on the Arecibo observatory and Green Bank telescope; the Parkes Pulsar Timing Array (PPTA) [8] based on the Parkes telescope; the European Pulsar Timing Array (EPTA) [9] based on five one-hundred meter class radio telescopes such as Effelsberg and Lovell. These collaborations share observational data and analysis tools within the framework of the International Pulsar Timing Array (IPTA) [10]. The IPTA has recently released timing data for 49 millisecond pulsars (MSPs) gathered over a period of 5 to 22 years [11].

The sensitivities of PTAs to stochastic GW background [12; 13], continuous waves from individual sources [14] and bursts with memory [15] have constantly improved over the past five years. The first PTA-based GW detection, possibly a stochastic background signal, is anticipated in the next a few years.

Next generation large-scale radio telescopes, such as the Five-hundred-meter Aperture Spherical Telescope (FAST) [16] and the Square Kilometer Array (SKA) [17], will bring profound

Content from this work may be used under the terms of the Creative Commons Attribution 3.0 licence. Any further distribution of this work must maintain attribution to the author(s) and the title of the work, journal citation and DOI.
Published under licence by IOP Publishing Ltd

changes and challenges in the enterprise of PTA by discovering hundreds of new stable MSPs and substantially improving timing precision.

Pulsar timing measures the arrival time of a fiducial phase point (usually the peak) on the integrated pulse profile which is obtained by folding the individual pulses across integration time and (after dedispersion) radio frequency channels. At 100 ns timing precision level, the main contributors in the error budget of the TOA measurement are the pulse phase jitter noise (due to the fluctuation in the shape and arrival time of individual pulses) and the additive radiometer noise (due to sky background and instrumental thermal electron noise) that can be estimated respectively by the following formulae [18; 19]

$$\sigma_{\mathrm{j}} \approx 0.28W\sqrt{\frac{P}{\Delta t}}\,, \tag{1}$$

$$\sigma_{\mathrm{r}} \approx \frac{WS}{F\sqrt{2\Delta f \Delta t}}\sqrt{\frac{W}{P-W}}\,. \tag{2}$$

Here Δt is the integration time to obtain an integrated pulse profile, P is the rotation period of the pulsar, W is the effective width of the integrated pulse profile, F is the flux density, Δf is the bandwidth. $S = 2\eta k_B (A_{\mathrm{eff}}/T_{\mathrm{sys}})^{-1}$ is the system equivalent flux density (SEFD) [20], where $\eta \sim 1.0$ is the system efficiency factor, k_B is the Boltzmann's constant, T_{sys} is the system temperature, A_{eff} is the effective collecting area, and $A_{\mathrm{eff}}/T_{\mathrm{sys}}$ is called the telescope sensitivity. The signal-to-noise ratio (S/N) of integrated pulse profile is defined as the height of the pulse profile divided by σ_r. The total noise of TOA σ_t satisfies $\sigma_t^2 = \sigma_j^2 + \sigma_r^2$.

2. Next generation facilities

2.1. FAST

FAST is located in the karst depression area in southwest China. It is the largest filled-aperture single dish radio telescope in the world, having a reflector that is 500 m in diameter with $A_{\mathrm{eff}} = 67,000\ \mathrm{m}^2$ (equivalently an aperture of about 300 m diameter). The active surface and feed cabin suspension systems enable a zenith angle coverage of up to 40° which gives a declination range of 66° to −14°. Potentially, the zenith angle can be extended up to 60° with $A_{\mathrm{eff}} = 31,000\ \mathrm{m}^2$ (equivalently an aperture of 200 m diameter). First light of FAST was achieved in September 2016 and it is currently under testing and commissioning.

Several sets of receivers have been proposed for FAST to cover the frequency range from 70 MHz to 3 GHz [21]. Among them, a 19-beam L-band receiver with $T_{\mathrm{sys}} = 20$ K and bandwidth of 400 MHz will be used ultimately to conduct the pulsar survey and timing. Simulations based on pulsar population models and the designed system parameters of FAST show that more than 5000 new pulsars would be discovered by the 19-beam receiver, among which about 10% would be millisecond pulsars. This survey can be carried out efficiently with eight hours survey time each day for 200 days [22]. During the early science run when the active control and 19-beam receiver are not available, FAST can discover new pulsars in the drift scan mode [23; 24]; this has already yeilded eight new discoveries so far [25].

2.2. SKA

SKA is a global effort to build the world's largest radio telescope with an effective collecting area of more than one square kilometer formed out of two major arrays: (i) SKA-Mid (350 MHz – 14 GHz) consisting of up to two thousand 15-m parabolic antennas (dishes) most of which will be located in South Africa, and (ii) SKA-low (50 MHz – 350 MHz) consisting of up to one million dipole antennas located in western Australia. SKA-Mid is the relevant instrument for the purpose of high-precision pulsar timing, while SKA-Low will help in pulsar detection as well as studying the interstellar medium through which radio pulses propagate. The SKA can access the entire sky that is visible from its geographic latitude to an angle off zenith of at least 85°.

The construction of SKA will be divided into two consecutive phases, SKA1 and SKA2. The construction of SKA1 is scheduled to last from 2018 to 2023, with early science operation starting as soon as 2021. Depending on progress in technology demonstration through SKA1 and funding, the construction of SKA2 is planned to start in 2023 and finish around 2030. Once completed, SKA is intended to be operational for more than fifty years, the typical lifetime of major radio telescopes.

In SKA1, 500 stations with each containing about 250 low frequency dipole antennas will be constructed for SKA1-Low, while SKA-Mid will have about 200 dishes incorporating 64 already existing ones from one of the SKA precursors, MeerKat [26]. The sensitivity $A_{\rm eff}/T_{\rm sys} > 2000$ m^2 k^{-1} for SKA1-Low and $A_{\rm eff}/T_{\rm sys} > 1000$ m^2 k^{-1} for SKA1-Mid. $A_{\rm eff}/T_{\rm sys} > 10^4$ m^2 k^{-1} for SKA2-Mid which is mostly relevant for the SKA era PTA.

Study based on astrophysical models of the pulsar population in our Galaxy and SKA design parameters shows that SKA1 is expected to discover 9000 canonical pulsars and 1500 MSPs by a survey with 65 day on SKA1-Low and 130 days on SKA1-Mid. For SKA2, up to 14000 canonical and 6000 MSPs can be discovered [17]. Due to their high intrinsic rotational stability, combined with the improved sensitivity of SKA, a timing uncertainty of < 100 ns is likely for a substantial fraction of the MSPs [27; 28].

3. Large-scale PTAs

In the following, we estimate the properties that a PTA can be expected to have in the FAST and SKA era. We will focus here on SMBHB sources in the frequency band $10^{-9} - 10^{-6}$ Hz. The Nyquist rate for the timing data, corresponding to twice the upper end of this frequency range, is one data point every two weeks.

3.1. *Pulsar timing*

For FAST, Smits et al. [22] found that it will take about 24 hours to time all of the 50 brightest MSPs once in order to obtain an integrated pulse profile for each pulsar at S/N of 500 that is required for high-precision timing. This estimate assumes a five minutes of integration time for pulse profile stabilization.

It is interesting to see what the make up of a future large-scale PTA may look like if the restriction on telescope time allocation for pulsar timing were removed. One can then work backwards and deduce how a given telescope time allocation will impact the science that can be done. In this spirit, we take Eq. 1 and Eq. 2 along with the design parameters of FAST and SKA to estimate the distribution of timing noise rms levels for the MSP population simulated by Smits et al. [17; 22]. Fig. 1 shows the number of MSPs having timing noise rms $\sigma_t \leq \alpha$, where $\alpha \in \{50, 100, 200, 500\}$ ns, one can expect to see with FAST or SKA given a certain maximum integration time per pulsar. (Note that this is a gross simplification of the actual situation where the integration time will depend on the pulsar brightness.) Specifically, we see that with FAST, 72 (99, 115) MSPs can be timed to be 100 ns or better with 10 (20, 30) mins integration time over an observation period of 0.3 (0.6, 1.0) day.

Combining high sensitivity and large field of view (FoV, 20 deg^2–250 deg^2), SKA can time MSPs more efficiently in time. Smits et al. [17] found that the observation time for timing 250 MSPs once with the best S/N (≈ 100) for the dense aperture array (AA) configuration will be 6 hours. This time can inflate to 15 hours for phased array feed 15-m dishes and 20 hours for single-pixel feed 15-m dishes. As before, we can read off from Fig. 1 that, for example, 189 (387, 557) MSPs can be timed to be 100 ns or better with 5 (10, 15) mins integration time, which takes 0.7 (1.8, 3.3) day. For SKA, the pulse phase jitter noise will dominate the radiometer noise for most of bright MSPs due to the high telescope sensitivity. In this case, the limiting

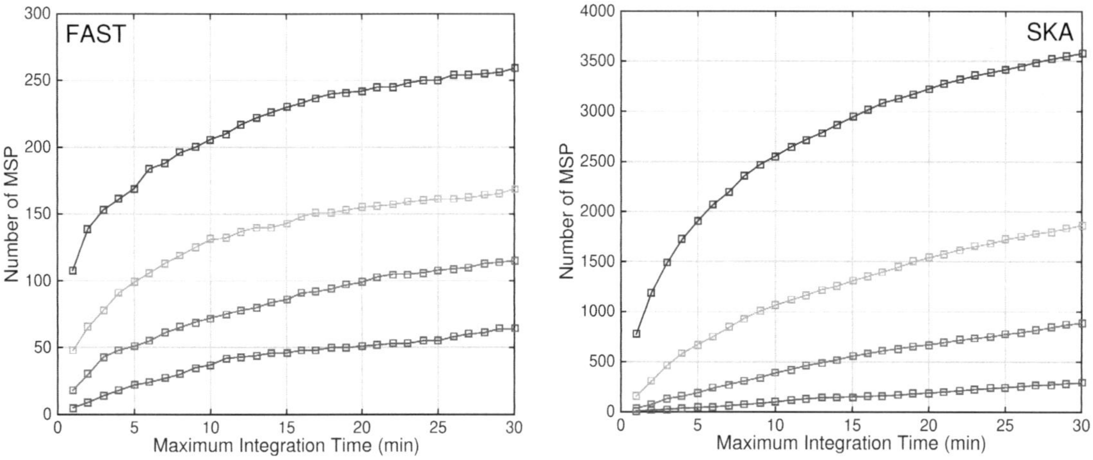

Figure 1. Number of MSPs having timing noise rms $\sigma_t \leq \alpha$ with varying maximum integration time. Here $\alpha \in \{50, 100, 200, 500\}$ ns, which is indicated by the color lines (from bottom to top): blue (50 ns), red (100 ns), yellow (200 ns), and purple (500 ns).

factor governing timing precision is the integration time. Therefore, an optimized way to operate telescope, e.g., observing multiple MSPs simultaneously [29], can be implemented to reduce the aforementioned observation time.

3.2. Single continuous wave source

The non-detection of a stochastic signal from the SMBHB population in the most sensitive search carried out till date [12] indicates that SMBHBs are sparser than anticipated by population synthesis models [30]. This may increase the likelihood of isolated SMBHB sources to be the first to be detected by PTAs. Based on simulations of a cosmological population SMBHBs, Rosado et al. [31] found that the probability of detecting single sources within 10 years operation of SKA1 will be 40% considering this probability will increase to 80% for SKA2. In a different work, Rosado et al. [32] emphasize the role of redshift bias in the detection of massive black hole system and how neglecting it underestimates the distance reach of current and future PTAs. Wang and Mohanty [33] conducted a realistic investigation, in the context of individual SMBHB sources, of the performance of a SKA-era PTA comprised of 10^3 MSPs selected from the simulated pulsar population in [17]. It was found that a SKA era PTA can significantly increase the distance reach to SMBHBs. For example, an all sky search can detect systems with redshifted chirp mass of 10^9 $M_\odot$ to redshift $z \sim 1$ and redshifted chirp mass of 10^{10} $M_\odot$ to redshift $z > 20$.

It is interesting to consider the possibility of merging detections of SMBHBs by a SKA-era PTA with ongoing and future searches for these systems, through luminosity variations in quasars, in the optical. Simultaneous observations in GW and optical can teach us much about the accretion physics in SMBHB systems by correlating the GW signal with optical variability. Typical error region for bright GW sources can be expected to be ~ 100 deg^2 [33], which can contain a large number of variable objects. However, the frequency of optical variability could be linked strongly to the (highly accurate) measured frequency of the GW signal and may help in narrowing down the optical counterpart. The depth of the search also implies that SMBHB candidates identified in the optical, such as PG 1302-102 [34], can be followed up in GWs. A detection of the GW signal from a candidate will provide the smoking gun evidence of its true nature.

3.3. Resolving multiple sources

The focus in current PTA-based GW searches has traditionally been on detecting a Gaussian, isotropic stochastic signal. However, the true signal is likely to be more complex [35; 36], and the multiple source problem [37] is getting more recognition in the PTA community. So far, studies [37; 38] of multiple source detection have assumed a simplified model in which the signals are embedded in white noise, the so-called pulsar term is dropped from the waveform of each, and there is no stochastic GW signal. Under these simplifications, it was shown that the simultaneous fitting of multiple SMBHB signals can be accomplished using a Genetic Algorithm. For a realistic assessment of a large-scale PTA, further development of data analysis methods that can work without these simplifying assumptions is required.

The high redshift reach of an SKA era PTA based search for SMBHBs implies that multiple resolvable systems may be detectable in the PTA data. Hence, the data analysis methods for future large-scales PTAs should not only be capable of handling $O(10^3)$ pulsars but also multiple sources and signal types, ranging from isolated sources to the stochastic signal from an unresolved population.

Acknowledgments

Y.W. is supported by the National Natural Science Foundation of China under grants 11503007, 91636111 and 11690021. The contribution of S.D.M. to this paper is partially supported by U.S. National Science Foundation grant PHY-1505861.

References

[1] Abbott B P, Abbott R, Abbott T D, Abernathy M R, Acernese F, Ackley K, Adams C, Adams T, Addesso P, Adhikari R X and et al 2016 *Physical Review Letters* **116** 061102 (*Preprint* 1602.03837)

[2] Amaro-Seoane P, Audley H, Babak S and et al 2017 *ArXiv e-prints* (*Preprint* 1702.00786)

[3] Luo J, Chen L S, Duan H Z and et al 2016 *Classical and Quantum Gravity* **33** 035010 (*Preprint* 1512.02076)

[4] Ölmez S, Mandic V and Siemens X 2010 *Phys. Rev. D* **81** 104028 (*Preprint* 1004.0890)

[5] Starobinskiĭ A A 1979 *Soviet Journal of Experimental and Theoretical Physics Letters* **30** 682

[6] Caprini C, Durrer R and Siemens X 2010 *Phys. Rev. D* **82** 063511 (*Preprint* 1007.1218)

[7] Demorest P B, Ferdman R D, Gonzalez M E and et al 2013 *Astrophysical Journal* **762** 94 (*Preprint* 1201.6641)

[8] Manchester R N, Hobbs G, Bailes M and et al 2013 *Publications of the Astronomical Society of Australia* **30** e017 (*Preprint* 1210.6130)

[9] Ferdman R D, van Haasteren R, Bassa C G and et al 2010 *Classical and Quantum Gravity* **27** 084014 (*Preprint* 1003.3405)

[10] Manchester R N 2013 *Classical and Quantum Gravity* **30** 224010 (*Preprint* 1309.7392)

[11] Verbiest J P W, Lentati L, Hobbs G and et al 2016 *Monthly Notices of the Royal Astronomical Society* **458** 1267–1288 (*Preprint* 1602.03640)

[12] Shannon R M, Ravi V, Lentati L T and et al 2015 *Science* **349** 1522–1525 (*Preprint* 1509.07320)

[13] Arzoumanian Z, Brazier A, Burke-Spolaor S and et al 2016 *Astrophysical Journal* **821** 13 (*Preprint* 1508.03024)

[14] Arzoumanian Z, Brazier A, Burke-Spolaor S and et al 2014 *Astrophysical Journal* **794** 141 (*Preprint* 1404.1267)

[15] Arzoumanian Z, Brazier A, Burke-Spolaor S and et al 2015 *Astrophysical Journal* **810** 150 (*Preprint* 1501.05343)

[16] Hobbs G, Dai S, Manchester R N, Shannon R M, Kerr M, Lee K J and Xu R 2014 *ArXiv e-prints* (*Preprint* 1407.0435)

[17] Smits R, Kramer M, Stappers B, Lorimer D R, Cordes J and Faulkner A 2009 *Astronomy and Astrophysics* **493** 1161–1170 (*Preprint* 0811.0211)

[18] Cordes J M and Shannon R M 2010 *ArXiv e-prints* (*Preprint* 1010.3785)

[19] Wang Y 2015 *Journal of Physics Conference Series* **610** 012019 (*Preprint* 1505.00402)

[20] Wilson T L, Rohlfs K and Hüttemeister S 2013 *Tools of Radio Astronomy*

[21] Nan R, Li D, Jin C, Wang Q, Zhu L, Zhu W, Zhang H, Yue Y and Qian L 2011 *International Journal of Modern Physics D* **20** 989–1024 (*Preprint* 1105.3794)

[22] Smits R, Lorimer D R, Kramer M, Manchester R, Stappers B, Jin C J, Nan R D and Li D 2009 *Astronomy and Astrophysics* **505** 919–926 (*Preprint* 0908.1689)

[23] Yue Y, Li D and Nan R 2013 *Neutron Stars and Pulsars: Challenges and Opportunities after 80 years* (*IAU Symposium* vol 291) ed van Leeuwen J pp 577–579 (*Preprint* 1211.0748)

[24] Zhang L, Hobbs G, Li D, Lorimer D, Zhang J, Yu M, Yue Y L, Wang P, Pan Z C and Dai S 2016 *Research in Astronomy and Astrophysics* **16** 151

[25] http://crafts.bao.ac.cn/pulsar/ accessed: 2017-10-31

[26] Brederode L R, van den Heever L, Esterhuyse W and Jonas J L 2016 *Ground-based and Airborne Telescopes VI* (*Proceedings of the SPIE* vol 9906) p 990625

[27] Kramer M, Backer D C, Cordes J M, Lazio T J W, Stappers B W and Johnston S 2004 *New Astronomy Reviews* **48** 993–1002 (*Preprint* astro-ph/0409379)

[28] Manchester R N 2010 *ArXiv e-prints* (*Preprint* 1004.3602)

[29] Lazio T J W 2013 *Classical and Quantum Gravity* **30** 224011

[30] Sesana A, Vecchio A and Volonteri M 2009 *Monthly Notices of the Royal Astronomical Society* **394** 2255–2265 (*Preprint* 0809.3412)

[31] Rosado P A, Sesana A and Gair J 2015 *Monthly Notices of the Royal Astronomical Society* **451** 2417–2433 (*Preprint* 1503.04803)

[32] Rosado P A, Lasky P D, Thrane E, Zhu X, Mandel I and Sesana A 2016 *Physical Review Letters* **116** 101102 (*Preprint* 1512.04950)

[33] Wang Y and Mohanty S D 2017 *Physical Review Letters* **118** 151104 (*Preprint* 1611.09440)

[34] Graham M J, Djorgovski S G, Stern D, Glikman E, Drake A J, Mahabal A A, Donalek C, Larson S and Christensen E 2015 *Nature* **518** 74–76 (*Preprint* 1501.01375)

[35] Sesana A, Vecchio A and Colacino C N 2008 *Monthly Notices of the Royal Astronomical Society* **390** 192–209 (*Preprint* 0804.4476)

[36] Kocsis B and Sesana A 2011 *Monthly Notices of the Royal Astronomical Society* **411** 1467–1479 (*Preprint* 1002.0584)

[37] Babak S and Sesana A 2012 *Phys. Rev. D* **85** 044034 (*Preprint* 1112.1075)

[38] Petiteau A, Babak S, Sesana A and de Araújo M 2013 *Phys. Rev. D* **87** 064036 (*Preprint* 1210.2396)

IOP Conf. Series: Journal of Physics: Conf. Series **957** (2018) 012004 doi:10.1088/1742-6596/957/1/012004

Identification and mitigation of Advanced LIGO noise sources

Beverly K. Berger
for the LIGO Scientific Collaboration
LIGO, California Institute of Technology, Pasadena, California 91125, USA
E-mail: beverly.berger@ligo.org

Abstract. In order to increase the reach of the astrophysical searches, various sources of instrumental and environmental noise must be identified and ameliorated. Here we discuss efforts to understand the origin of noise manifested as short-duration bursts (glitches) and/or range-impacting features at LIGO Hanford. Several examples found at LIGO Hanford Observatory in O1 and O2 were identified including glitches due to an air compressor, ringing phone, airplanes, and an incorrect servo setting, and a decrease in detector sensitivity due to truck traffic.

1. Introduction

GW150914, the first direct detection of gravitational waves from a binary black hole merger was a milestone for physics and astronomy [1]. Making this detection required instruments of exquisite sensitivity that could measure a length change equivalent to the width of an atom in the distance from the earth to the sun. To achieve this sensitivity, the characteristics of the noise in the Advanced LIGO (aLIGO) instruments [2] must be studied, leading to improved sensitivity and improved ability to distinguish signal from noise. The motivation to study the noise is clear. Glitches are transient noise events that could mimic astrophysical events [3] or indicate badly behaving components of the detectors. Classes of glitches with similar properties offer a target for identification of their cause. Even if the cause of a class of glitches is unknown, glitches known to belong to a class can be identified and removed from any influence on the data analysis [4, 5]. So far, the aLIGO detectors have engaged in two observing runs, O1 and O2, taking data from 12 September 2015 to 19 January 2016 and from 30 November 2016 to 25 August 2017 respectively.

aLIGO noise is non-Gaussian and non-stationary.[1] This means that from day to day or even hour to hour both the overall level of noise and the character of that noise can change. For this reason, the LIGO Scientific Collaboration (LSC) initiated a program of data quality (DQ) shifts wherein a collaboration member is assigned to monitor remotely either the detector at Hanford (LHO) or at Livingston (LLO) for several days and then report back to the Detector Characterization (DetChar) group on any changes or issues noticed during the DQ shift that might affect the astrophysical searches. Urgent issues are reported directly to staff at the site as soon as possible. Most DQ shift reports are summarized in the public aLIGO logbooks [6, 7].

[1] Most of the time, the noise is sufficiently Gaussian and stationary for astrophysical events to be identified (see the discussion in, e.g., [8]) but care must be taken.

Content from this work may be used under the terms of the Creative Commons Attribution 3.0 licence. Any further distribution of this work must maintain attribution to the author(s) and the title of the work, journal citation and DOI.
Published under licence by IOP Publishing Ltd

Tools developed by the DetChar group members make it possible to keep a close eye on the instruments. Key to the ability to perform remote DQ shifts are the Summary Pages[2] (SP) [9]. For those in the LSC, the SP provide information such as time series, spectrograms, spectra, and glitch activity of the gravitational wave channel $h(t)$ and many other channels of these complex instruments. They are available with about one-hour latency through any internet connection.

Three additional tools have played a role in the work described here. Omicron [11] is a software algorithm that identifies glitches in the data according to features such as peak frequency , signal-to-noise ratio (SNR), and bandwidth. The SP include glitchgrams, Omicron's output in terms of peak frequency vs time, color coded for SNR. Advanced LIGO monitors thousands of channels in addition to the gravitational wave channel $h(t)$. These auxiliary channels can be studied to find correlations between instrumental or environmental sensor signals and the gravitational wave channel. To make use of this information, a further analysis is carried out by hVeto [12] which identifies statistically significant correlations between glitches in $h(t)$ and glitches in auxiliary channels. hVeto is applied hierarchically to identify independent glitch families associated with different sets of auxiliary channels. Finally, a sophisticated plotting package, ligoDV-web (ldvw) [13], allows exploration of the data in all channels.

In the remainder of this paper, we will discuss several examples including the 50 Hz glitches traced to an air compressor, the 1083 Hz glitches traced to an incorrectly set servo gain, peculiar features in the $h(t)$ spectrogram found to be caused by a ringing phone, airplane glitches, and the 5 am Pacific-time truck-traffic range drop. The focus will be on discovery of the class of glitches and steps taken to understand their nature, and, if possible, fix their cause. We will only consider glitches found at LHO.

2. Examples

The 50 Hz glitches. The first example is a series of glitches seen early in O1 in the gravitational wave channel, $h(t)$, with peak frequency of 50 Hz visible in Fig. 1.

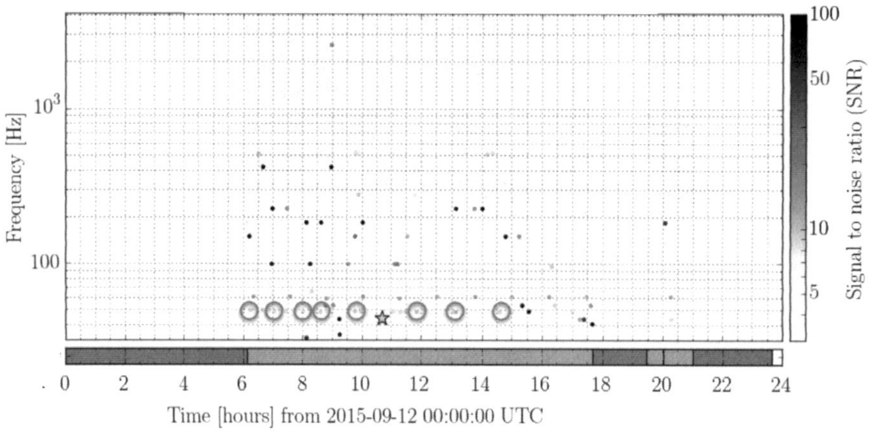

Figure 1. The 50 Hz glitches. A glitchgram is shown for the entire day of 12 September 2015 with time on the horizontal axis and frequency on the vertical axis. The glitches are color coded by SNR using yellow-to-black with yellow indicating lower SNR and black indicating higher SNR. The 50 Hz glitches, circled, are about 1 hour apart.

[2] Note that a version of the Summary Pages has recently become publicly available on the LIGO Open Science Center website [10].

When analyzed by hVeto, the most significant channel was a channel measuring the light incident on a photodiode used for alignment control of the X-arm of the LHO detector. This was not surprising. However, the display of all significant channels showed that environmental channels recording seismic vibrations at the end of the X-arm of the detector were also correlated with the glitches in $h(t)$. This meant that there was a relationship between the 50 Hz glitches and ground vibrations at the X end station. The DetChar group then suggested [14] that the glitching in $h(t)$ originated in the electric power mains but were coupling to the gravitational wave channel through a seismic connection. It was then speculated that the culprit was a chiller compressor cycling on and off in response to the temperature in its immediate environment. Sure enough [15], when LHO staff visited the end station on 30 September 2015 to look for the problem, they discovered that the vibration-isolation feet on the air compressor at the end station were ineffective and needed to be replaced. They were replaced and the 50 Hz glitches disappeared. While other glitch families were identified and mitigated in O1, the remaining examples presented here occurred in O2.

The 1083 Hz glitches. The first of these were noticed early in O2 as a line of glitches in the daily gltichgram centered on 1083 Hz. Detailed spectrograms showed continual glitching at frequencies around the line. This particular frequency corresponds to one of the LHO calibration (spectral) lines [16] used to determine the actual value of the observed strain. It was first supposed that the problem might arise from fluctuations in the calibration line. However, that was ruled out when the line was briefly turned off. The gltiching did not change. This is shown

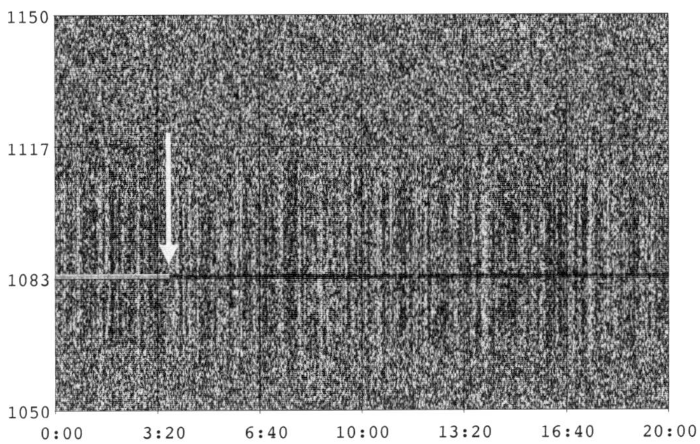

Figure 2. The glitching around the 1083 Hz calibration line. A 20-minute spectrogram of $h(t)$ starting at 21:30 UTC on 15 November 2016 is shown. The yellow arrow points to the time at which the calibration line is turned off. The vertical yellowish bands arise from nearly continual glitching and persist after the line is turned off. The horizontal axis shows the time in minutes after the start time while the vertical axis shows frequency in Hz.

in Fig. 2. For some weeks, these glitches remained a puzzle. hVeto often reported correlations with some channels but the alignment in time was not precise. Apparently, because $h(t)$ was glitching all the time, it accidentally correlated with other channels that were glitching all the time. Finally, a breakthrough occurred when the first appearance of the glitching was tracked down to the second lock stretch on 11 October 2016. What happened at that time? It turned out that output mode cleaner (OMC) length-dither-line amplitudes had been lowered on that day. There should have been an accompanying OMC servo gain to compensate for the change but that did not happen (although no one realized it at the time). On 9 January 2017, the servo gain was reset to the correct value and the gltiching disappeared [17].[3] In fact, the channel

[3] Initially, the correction was temporary to see if the problem had been identified. There was a concern that fixing the glitch problem might create other issues. When that did not to happen, the change was made permanent.

IOP Conf. Series: Journal of Physics: Conf. Series **957** (2018) 012004 doi:10.1088/1742-6596/957/1/012004

associated with the OMC length dither was glitching strongly at frequencies of 4 – 7 kHz and these glitches did align with those in $h(t)$ around 1083 Hz.

The phone rang. On 4 December 2016 at about 17:45 UTC, one of the LHO operators noticed a peculiar series of glitches. This was then independently noticed by offsite DQ experts. A spectrogram of the time interval of the peculiar glitching is shown in Fig. 3. After some head-scratching, someone thought to create a sound file of the spectrogram. Note that this is meaningful because the aLIGO frequencies are in the range of human hearing. The sound file revealed a loudly ringing phone, followed by a softer ring, followed by an answering machine and some talking before another loud ring was heard. Oops! During science runs, there are not supposed to be ringing phones near the vacuum enclosure of the instrument. Of course this problem was fixed by making sure all phones were either removed or turned off [18].

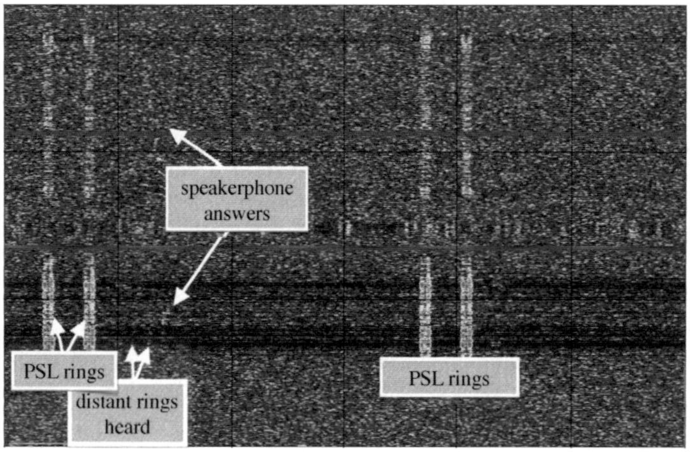

Figure 3. Spectrogram of strange glitching in $h(t)$. A pattern of regular broad-band glitches with some "scruff" in between is shown along with indicators of the eventual explanation. The phone creating the problem was located in the pre-stabilized laser (PSL) chamber.

Airplane overhead. One source of noise for LHO in initial LIGO was caused by airplane flyovers through acoustic coupling to the buildings housing the instruments. In fact, a software monitor had been developed to track plane flights over the site by tracking Doppler shifts picked up by microphones at the X- and Y-arm end stations and at the connecting corner station as the plane moves over the various parts of the instrument [19]. This was not expected to be an issue for aLIGO due to its greatly enhanced seismic isolation. However, on 1 March 2017 at about 1:51 UTC, a peculiar noise pattern was seen in the spectrogram of $h(t)$ with a duration of more than a minute with two parallel bands with frequencies decreasing in time from about 100 Hz to about 70 Hz. hVeto reported the involvement of microphone channels. Fig. 4 shows an overlay of $h(t)$ on the spectrograms of microphone channels at the X-end (EX), Y-end (EY), and corner station (CS). The noise in $h(t)$ follows whichever microphone signal is loudest at the time as the sound from the plane couples to the nearest interferometer mirrors [20]. After this initial airplane identification, additional evidence of airplane glitches were seen. While the correlation of these glitches with microphone channels means that they cannot be mistaken for astrophysical events, their elimination from $h(t)$, ideally by reducing the coupling, is a high priority for future work [21].

The noon UTC truck noise. It was noticed that almost every week, Monday through Thursday, the detector sensitivity as measured by the distance to which a binary neutron star inspiral could be detected (BNS range) showed a dip at about the same time every day. The range dip corresponded to an increase in seismic noise as measured in the corner station (but not in end-X or end-Y) in the 10 – 30 Hz band as is shown in Fig. 5. Further inspection showed a second dip at about 0:00 UTC on the same days. The local times at LHO were 5 am and 5 pm PDT respectively. In the previous month, before standard time (PST) changed to daylight-

IOP Conf. Series: Journal of Physics: Conf. Series **957** (2018) 012004 doi:10.1088/1742-6596/957/1/012004

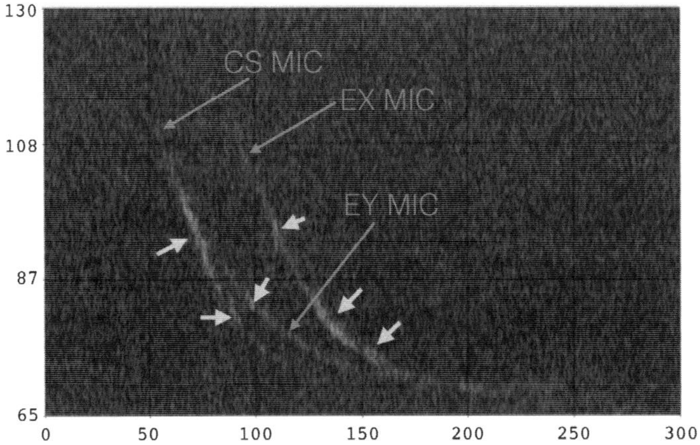

Figure 4. Airplane flyover glitch. This spectrogram overlays $h(t)$ and 3 microphone channels for time (horizontal axis) running for 300 seconds after 1 March 2017 01:50 UTC and frequencies (vertical axis) running from 65 Hz to 130 Hz on a linear scale. The yellow arrows point to bright regions in the display from the overlay of $h(t)$. The long arcs are the traces from the microphones located at the end of the X-arm (EX MIC) and the Y-arm (EY MIC) and at the corner station (CS MIC).

saving time (PDT), the same effect occurred at the same local times (5 am and 5 pm PST) but one hour later in UTC. Previously, it had been suggested that the culprit was truck traffic passing close to the corner station on the road that passes the LHO entrance. The observed behavior reinforced this conclusion. It was recognized that there was strong coupling of 8 – 18 Hz input beam vibration to the 70 – 200 Hz band of $h(t)$ which would account for the effect due to the truck traffic. It was suggested that the coupling was caused by a baffle in the input arm of the interferometer with several openings for various beams, nicknamed the Swiss Cheese Baffle [24], and could be reduced by adding dampers to the baffle. In May 2017, O2 was interrupted to make some repairs at both sites. The dampers were added at that time.[22, 23, 24] After the dampers were added, the seismic increase was still present but the range drop had disappeared.

3. Final thoughts

These are some examples of identifying features in $h(t)$ associated with particular types of noise. In some cases, the existence of the noise can be traced to a particular problem that can be fixed. Sometimes, even if the cause is known, the problem cannot be easily fixed. However, the now understood character of the noise can be used to avoid confusing it with a signal. While the 50 Hz glitches, phone, and airplane events would not have been misinterpreted as astrophysical events, they could mask actual signals. The 1083 Hz glitches interfered with with a search for unmodeled astrophysical events. The truck-caused range drop made the instrument less sensitive for a significant amount of time and thus less likely to detect a signal during that time. Thus understanding and mitigating these noise sources is a high priority for astrophysical event detection. For more on this topic see [25].

Acknowledgments

I would like to thank S. Dwyer, J. McIver, B. O'Reily, and J. Smith for valuable comments. LIGO was constructed by the California Institute of Technology and Massachusetts Institute of Technology with funding from the National Science Foundation and operates under cooperative agreement PHY-0757058. This paper carries the LIGO Document Number P1700343.

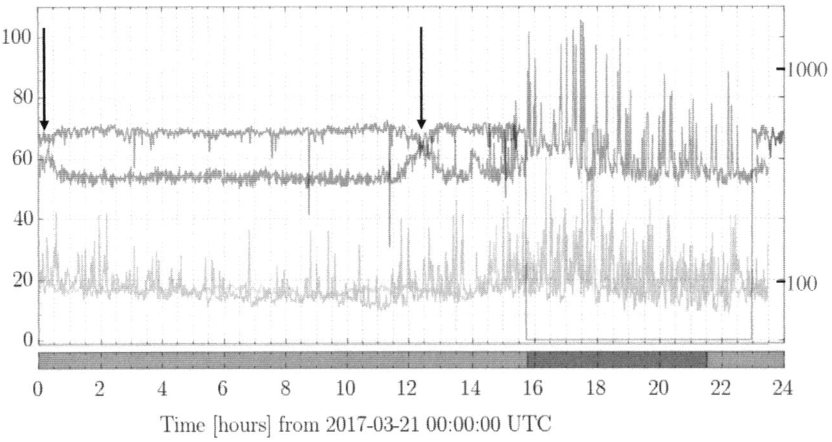

Figure 5. Truck noise at 12:00 UTC. The range dip and excess seismic noise are shown for the entire day of 21 March 2017. There is a corresponding dip and excess at about 0:00 UTC. Both times are marked by arrows. The BNS range is shown in red (left axis in Mpc) and the seismic noise (ground motion) in the 10 – 30 Hz band at the corner station is shown in gray (right axis in nanometers per second). The remaining traces show seismic motion at the end stations.

References

[1] Abbott, B.P. et al 2016 Phys. Rev. Lett. **116**, 061102
[2] Aasi, J. et al 2015 Class. Quantum Grav. **32** 074001
[3] Abbott, B.P. et al, arXiv:1710.02185 [gr-qc]
[4] Aasi, J. et al 2015 Class. Quantum Grav. **32** 105012
[5] Abbott, B.P. et al 2016 Class.Quant.Grav. 33, 134001
[6] aLIGO LHO Logbook, https://alog.ligo-wa.caltech.edu/aLOG/
[7] aLIGO LLO Logbook, https://alog.ligo-la.caltech.edu/aLOG/
[8] Abbott, B.P. et al 2017 Phys. Rev. Lett. **119** 141101
[9] https://github.com/gwpy/gwsumm/blob/master/docs/overview.rst
[10] LIGO Open Science Center, https://losc.ligo.org
[11] Robinet, F. 2015 Omicron: An Algorithm to Detect and Characterize Transient Noise in Gravitational-Wave Detectors. https://tds.ego-gw.it/ql/?c=10651
[12] Smith, J.R. et al 2011 Class. Quantum Grav. **28** 235005
[13] Areeda, J.S. et a; 2017 Astronomy and Computing **18** 2734 (2017)
[14] https://alog.ligo-wa.caltech.edu/aLOG/index.php?callRep=21436
[15] https://alog.ligo-wa.caltech.edu/aLOG/index.php?callRep=22119
[16] Cahillane, C. arXiv:1708.03023 [astro-ph.IM]
[17] https://alog.ligo-wa.caltech.edu/aLOG/index.php?callRep=33104
[18] https://alog.ligo-wa.caltech.edu/aLOG/index.php?callRep=32503
[19] Goetz, E. and Riles, K. PlaneMon: Airplane Detection Monitor, https://alog.ligo-wa.caltech.edu/aLOG/index.php?callRep=33104
[20] https://alog.ligo-wa.caltech.edu/aLOG/index.php?callRep=34547
[21] https://dcc.ligo.org/LIGO-T1700214/public, p. 74.
[22] https://alog.ligo-wa.caltech.edu/aLOG/index.php?callRep=35166
[23] https://alog.ligo-wa.caltech.edu/aLOG/index.php?callRep=35735
[24] https://alog.ligo-wa.caltech.edu/aLOG/index.php?callRep=36147
[25] Nuttall, L.K., "Characterising transient noise in the LIGO detectors," in preparation

Gravitational Waves Propagation through the Stochastic Background of Gravitational Waves

Frajuca, C.[1], Bortoli, F.S.[1], Nakamoto, F.Y.[1] and Santos, G.A.[1]
[1] Sao Paulo Federal Institute – Sao Paulo, SP, Brazil

E-mail: Frajuca@gmail.com

Abstract: With the recent claim that gravitational waves were finally detected and with other efforts around the world for GWs detection, its is reasonable to imagine that the relic gravitational wave background could be detected in some time in the future and with such information gather some hints about the origin of the universe. But, it's also be considered that gravity has self-interaction, with such assumption it's reasonable to expect that these gravitational wave will interact with the relic or non-relic GW background by scattering, for example. Such interaction should decrease the distance which such propagating waves could be detected The propagation of gravitational waves (GWs) is analyzed in an asymptotically de Sitter space by the perturbation expansion around Minkowski space using a scalar component. Using the case of de Sitter inflationary phase scenario, the perturbation propagates through a FRW background. The GW, using the actual value for the Hubble scale (H_0), has a damping factor with a very small valor for the size of the observational universe; the stochastic relic GW background is given by a dimensionless function of the frequency. In this work we analyze this same damping including the gravitational wave background due to astrophysical sources such background is 3 orders of magnitude bigger in some frequencies and produces a higher damping factor.

1. Introduction

The detection of gravitational waves came after a long road of experiments planned in 2010 [1], in 2016 finally the detection was made [2,3]. Gravitational waves got a very strong evidence with the **PSR B1913+16** (also known as **PSR J1915+1606**, **PSR 1913+16**, and the **Hulse–Taylor binary** after its discoverers) is a pulsar (a radiating neutron star) which together with another neutron star thus forming a binary star system. PSR 1913+16 was the first binary pulsar to be discovered and its orbital period is decreasing with time due the emission of gravitational waves [4]. The first attempts to gravitational wave detection starts in the early sixties [5] with the resonant mass gravitational wave detection [6,7,8,9]. The efforts towards the detection of gravitational waves using this kind of detector culminated with the spherical detector where six sensors are connected to the surface of the sphere, arranged according to a work based on the distribution of Merkowitz and Johnson [10,11], these transducers are located as if they were in the center of 6 pentagons connected in a surface corresponding to half dodecahedron [12,13]. Each transducer mechanically amplifies the motion occurring at the detection rate of the detector in the region of the sphere in which it is connected. The already amplified movement excites the membrane of one resonant cavity.

The Brazilian efforts on the field can be seen in [14,15,16,17,18,19,20,21,22,23,24,25,26,27].

Content from this work may be used under the terms of the Creative Commons Attribution 3.0 licence. Any further distribution of this work must maintain attribution to the author(s) and the title of the work, journal citation and DOI.
Published under licence by IOP Publishing Ltd

The theory postulates that gravitons are always attractive (gravity never repels), acting beyond any distance (gravity is universal) and come from an unlimited number of objects. Therefore, if graviton exists, it must be a boson with spin equal to two, and must have a zero rest mass, according to Quantum Mechanics.

The gravitons were postulated after the great success of quantum theory (especially the Standard Model) in describing the behavior of all other forces known to nature as transmitted by elementary particles: electromagnetism through the photon, the strong nuclear force by gluons, and the weak nuclear force by the W and Z bosons. The hypothesis is that the gravitational force is similarly transmitted by an elementary particle, which has not yet been discovered yet, called graviton, rather than described in terms of space-time curvature as general relativity. At the classical limit, both approaches provide identical results, and compatible with Newton's law of gravitation

The granulation of quantum theory is not compatible with the uniformity of Einstein's general relativity. These problems, along with some conceptual puzzles, have led many physicists to believe that a more complete theory than general relativity should regulate behavior close to Planck's length. String theory finally emerged as the most promising solution, it is the only known theory in which the quantum correction of any order for graviton dispersion is finite, and is always stable. But it is totally conjectural because string theory is not validated through experiments since its range of energies is far beyond what our experimental ability allows.

For this reason it is tried in this work to obtain a form to obtain results similar to the one of graviton scattering however using the theory of general relativity with a scalar component to try to prove the validity of this method and in the future to apply the same method using only generative relativity.

Gravitational waves can be seen as the coherent state of many gravitons, with both electromagnetic waves being coherent states of photons

2. The scope of this work

With the detection of gravitational wave its is reasonable to imagine that the relic gravitational wave background could be detected in some time in the future and with such information gather some hints about the origin of the universe [28,29]. Some calculations in the amplitude of some background has been made, as can be seen on figure 1.

But, it's need to be considered that gravity has self-interaction, it means that graviton interacts with gravitons [30, 31] as can be seem in figure 2. With such assumption it's reasonable to expect that these gravitational wave will interact with the relic or non-relic GW background by scattering, making some of the gravitational wave that is traveling in space loses coherence and deviate in another direction. Such interaction should decrease the distance which such propagating waves could be detected or at least making the signal weaker what makes it appear that the source is far away.

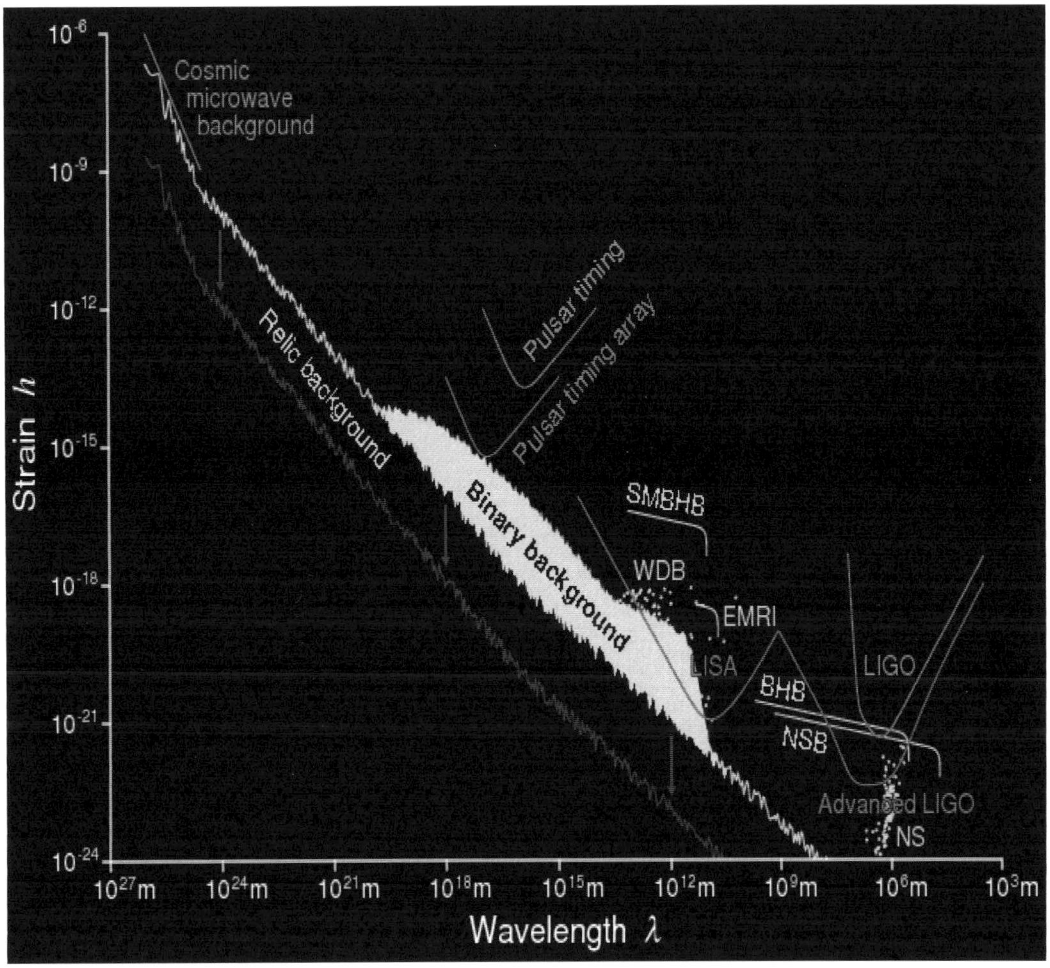

Figure 1. The gravitational wave background and the possible detector for each frequency band as can be found in the web page of Caltech http://www.tapir.caltech.edu/~teviet/Waves/gwave_spectrum.html, SMBHB means Supermassive Black Hole Binaries; WDB means White Dwarf Binaries; EMRI means Extreme Mass Ratio Inspirals and NSB means Neutron Star Binaries.

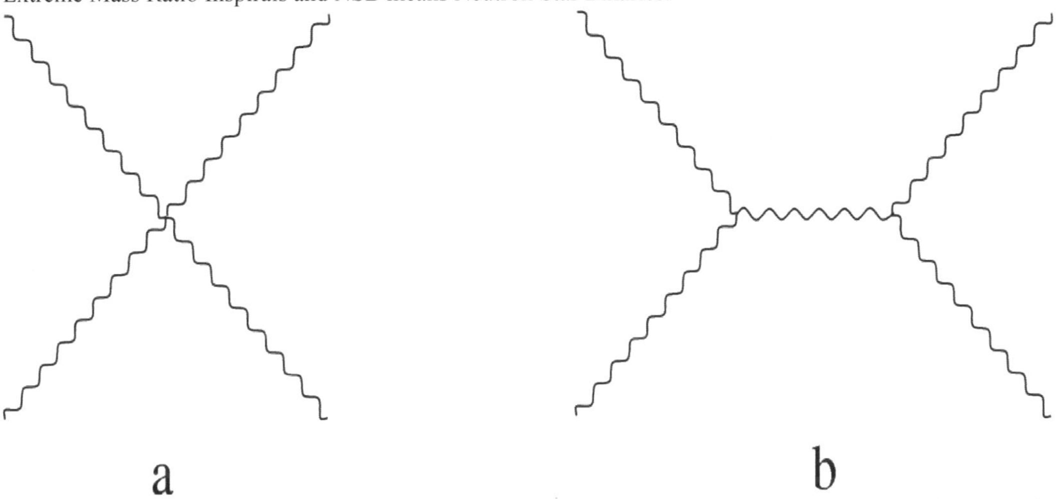

Figure 2. Four-graviton scattering in the first two orders of the perturbation theory (panels a and b, respectively).

IOP Conf. Series: Journal of Physics: Conf. Series **957** (2018) 012005 doi:10.1088/1742-6596/957/1/012005

The propagation of gravitational waves (GWs) is analyzed in an asymptotically de Sitter space by the perturbation expansion around Minkowski. The stochastic relic GW background is given by a dimensionless function of the frequency [28 and 29]. Another way to characterize such background could be seem in [32, 33]. Using the case of de Sitter inflationary phase scenario [28, 29, 34, 35], the perturbation propagates through a FRW background with a metric

$$ds^2 = a^2(\eta)\left(-d\eta^2 + dx^2\right) \tag{1}$$

where the negative term is the conformal time.

The GW is represented by

$$h_{\mu\nu} = e_{\mu\nu}\phi(\eta)\, e^{ik\cdot x} \tag{2}$$

The perturbation propagation metric could be expressed as

$$g_{\mu\nu} = a^2(\eta)\left(-d\eta^2 + dx^2 + h_{\mu\nu}dx^\mu dx^\nu\right) \tag{3}$$

The amplitude of the GW must satisfy the following equation:

$$\left(\frac{d^2}{d\eta^2} + \frac{2}{a}\frac{da}{d\eta}\frac{d}{d\eta} + |k|^2\right)\phi = 0 \tag{4}$$

the relevant result is the one corresponding to the de Sitter inflationary phase, where the scale factor behaves as

$$a \propto e^{H_{ds} t} \tag{5}$$

in terms of the physical time t. In this scenario the solution becomes

$$\phi(\eta) = \frac{a(\eta_1)}{a(\eta)}\left(1 + iH_{ds}\omega^{-1}\right)e^{-ik(\eta-\eta_1)} \tag{6}$$

here H_{ds} is the Hubble factor during the de Sitter phase.

It shows that, according to [35] the amplitude has a damping effect that goes with

$$\phi \propto e^{-H_{ds} t} \tag{7}$$

where H_{ds} is the Hubble parameter in the De Sitter phase. It shows that after a distance equal to $1/H_{ds}$ the amplitude of the wave decreases to a factor of 1/e.

Now if the actual value for the Hubble scale H_0, has a damping factor with a very small valor for the size of the observational universe. In this work we analyze this same damping including the gravitational wave background due to astrophysical sources. Such background is 3 orders of magnitude bigger in some frequencies and produces a considerable damping at shorter distances.

These results are presented in [34] and gave the authors the idea of developing this work, nevertheless it was done using the cosmological constant and not the GW background.

IOP Conf. Series: Journal of Physics: Conf. Series **957** (2018) 012005 doi:10.1088/1742-6596/957/1/012005

3. Results

This shows that the GW, using the actual value for the Hubble scale Ho, the amplitude of the GW decreases by a factor of $1/e$ after 100BLY propagating in the relic gravitational wave background.

What will happen if the gravitational background relic has the values that the GW background has today?

Figure 1 shows that the gravitational wave background has a amplitude bigger than the relic gravitational background by a factor of $10^{1.5}$ in strain h in all the wavelengths. The stochastic GW background is a energy density distribution ρ_{GW} , which is proportional to the GW strain squared (h^2), and the critical density to close the universe is given by:

$$\rho_C = \frac{3H_0^2}{8\pi G} \tag{8}$$

the gravitational wave relic is part of the critial density and depends in h^2 , as the authors want to increase value of the gravitational wave relic to the value of the gravitational wave background, let assume that the critical density increases by the same factor, then H_0^2 should increase by the same order of h^2.

This implies that the GW, using the actual value for the Hubble scale H_0, correct by a factor of $10^{1.5}$, the amplitude of the GW decreases by a factor of $1/e$ after about 3BLY.

4 Conclusions

Taking in consideration that gravitons should scatter as they interact with other gravitons, gravitational waves should do the same as they interact with the gravitational wave background. This effect is small but as gravitational waves travel very long distances the effect should become important.

This is just a preliminary result and much work is yet to be done.

As the gravitational wave measured in [1,2] was identified to be occurred in a distance of 1BLY without taking this effect into account, it could be located in a distance 30% closer, if the approximations used here are correct.

The next step to this work is to make the same calculations using only General Relativity without the use of scalar components.

Acknowledgements

Carlos Frajuca acknowledges FAPESP for grant #2013/26258-4 and grant #2006/56041-3.

References

[1] The Gravitational Waves International Committee Roadmap (GWIC). A global pan. 2010. Glasgow: University of Glasgow - Department of Physics and Astronomy - Kelvin Building (G12 8QQ), 117p.
[2] LIGO Scientific Collab. and Virgo Collab. (Abbott B P et al.) 2016 Phys.Rev. Lett., **116** (6): 061102
[3] Castelvecchi D, Witze A Nature News doi:10.1038/nature.2016.19361. Retrieved 11 February 2016[2] Taylor J H, Hulse R A, Fowler L A, Gullahorn G E, Rankin J M 1976 *Astrophysical Journal* **206** L53
[4] Taylor J H, Hulse R A, Fowler L A, Gullahorn G E, Rankin J M 1976 *Astrophysical Journal* **206** L53
[5] Weber J. 1960 *Physical Review* **117** 306
[6] Richard J P 1984 *Physical Review Letters* **167** 165
[7] Richard J P Proc. Second Marcel Grossmann Meeting on General Relativity, edited R.Ruffini, (North-Holland, Amsterdam).
[8] Frajuca C *et al* 2004 *Class. Quantum Grav.* **21** 1107
[9] Frossati G Proc. First International Workshop for an Omnidirectional Gravitational Radiation Observatory, W.F. Velloso, Jr., O.D. Aguiar and N.S. Magalhaes, editors (Singapore, World Scientific, 1997).
[10] Merkowitz S M and Johnson W W 1997 *Phys. Rev. D* **56** 7513
[11] S. M. Merkowitz and W. W. Johnson 1993 P*hys. Rev. Lett.* **70**, 2367
[12] Frajuca, C. "Otimização de transdutores de dois modos mecânicos para detectores de ondas gravitacionais." São Paulo. 97 p. PhD Dissertation - Universidade de Sao Paulo, 1996.
[13] Frajuca C. *et al* , Proc. 3rd Edoardo Amaldi Conference on Gravitational Waves (Pasadena, USA, July 1999). AIP Conf. Proc. 523 (New York, AIP) p.417

[14] Aguiar O D et all 2011 *Journal of Physics: Conference Series* **363** 012003
[15] O. D. Aguiar et. al 2005 *Class. Quantum Grav.* **22**, 209
[16] Frajuca *et al* 2002 *Class. Quantum Grav.* **19** 1961
[17] Magalhaes N S *et al* 1997 *Astrophysical Journal* **475**, 462
[18] Magalhaes N S *et al* 1995 *MNRAS* **274**, 670
[19] Frajuca C, Bortoli F S, Magalhaes N S 2005 *Brazilian Journal of Physics* **35** 1201
[20] Aguiar O D et al. 2006 *Journal Class. Quantum Grav.* **23**, 239
[21] Frajuca C, Bortoli F S, Magalhaes N S 2006 Journal of Physics: Conference Series **32** 319
[22] Frajuca C, Bortoli F S, Magalhães N S 2005 *Brazilian Journal of Physics* **35** 1201
[23] Aguiar O D et al., *Class. Quantum Grav.* **21** 459, 2004.
[24] Frajuca C, Magalhaes N S, Horiguti A M 2008 *Journal of Physics: Conference Series* **122** 012029
[25] Bortoli F S *et al* 2010 *Journal of Physics: Conference Series* **228** 012011.
[26] Magalhaes *et al* 1997 *Gen. Relat. Grav.* **29** 1511
[27] Magalhaes *et al* 1997 *Astroph. J.* **475** 462
[28] Allen B The stochastic gravity-wave background: Sources and Detection, in *Relativistic Gravitation and Gravitational Radiation* (Les Houches, 1995), p. 373
[29] Maggiore M 2000 *Phys. Rep.* **331** 283
[30] Tsvi P 2016 *International Journal of Modern Physics D* **25** 1644020
[31] Anirban B 2016 *CQG* **33** 125028
[32] Capozziello S, Corda C, De Laurentis M 2007 *Mod. Phys. Lett.* A **22** 2647
[33] Capozziello S, Corda C, De Laurentis M 2007 *Mod. Phys. Lett.* A **22** 1097
[34] Nowakowski M, Arraut I 2010 *Acta Physica Polonica B* **41** 911
[35] Arraut I 2013 *Modern Physics Lettes A* **28** 1350019

High-Reflection Coatings for Gravitational-Wave Detectors: *State of The Art and Future Developments*

Alex Amato[1], Gianpietro Cagnoli[1], Maurizio Canepa[2], Elodie Coillet[3], Jerome Degallaix[1], Vincent Dolique[1], Daniele Forest[1], Massimo Granata[1], Valérie Martinez[3], Christophe Michel[1], Laurent Pinard[1], Benoit Sassolas[1], Julien Teillon[1]

[1] *Laboratoire des Matériaux Avancés, CNRS/IN2P3, F-69622 Villeurbanne, France.*
[2] *OPTMATLAB, Università di Genova, Via Dodecaneso 33, 16146 Genova, Italy.*
[3] *Institut Lumière et Matière, CNRS, Université de Lyon, F-69622 Villeurbanne, France.*

E-mail: a.amato@lma.in2p3.fr

Abstract. We report on the optical, mechanical and structural characterization of the sputtered coating materials of Advanced LIGO, Advanced Virgo and KAGRA gravitational-waves detectors. We present the latest results of our research program aiming at decreasing coating thermal noise through doping, optimization of deposition parameters and post-deposition annealing. Finally, we propose sputtered Si_3N_4 as a candidate material for the mirrors of future detectors.

The high-reflecting (HR) coatings of the gravitational-wave (GW) detectors Advanced LIGO [1], Advanced Virgo [2] and KAGRA [3] have been deposited by the Laboratoire des Matériaux Avancés (LMA) in Lyon (Fr), where they have been the object of an extensive campaign of optical and mechanical characterization. In parallel, an intense research program is currently ongoing at the LMA, aiming at the development of low-thermal-noise optical coatings. The materials presented in this study are deposited by ion beam sputtering (IBS), using different coaters: a commercially available Veeco SPECTOR and the custom-developed DIBS and Grand Coater (GC). Unless specified otherwise, each coater uses different sets of parameters for the ion beam sources. Coating refractive index and thickness are measured by transmission spectrophotometry at LMA using fused silica substrates (⌀ 1", 6 mm thick) and by reflection spectroscopic ellipsometry [4] at the OPTMATLAB using silicon substrates (⌀ 2", 1 mm thick). Results of the two techniques are in agreement within 3% and here are presented the average values, used to calculate coating density. Structural properties are probed by Raman scattering at the Institut Lumière Matière (ILM), using fused silica substrates. Finally, coating loss angle ϕ_c is measured on a Gentle Nodal Suspension [5, 6] (GeNS) system at LMA, with disk-shaped resonators of fused-silica (⌀ 2" and 3" with flats, 1 mm thick) and of silicon (⌀ 3", 0.5 mm thick). ϕ_c is evaluated using the resonant method [7] i.e. by measuring the ring-down time of several vibrational modes of each sample. For the i-th mode, it writes

$$\phi_{i,c} = \frac{1}{D_i}\left[\phi_{i,\text{tot}} - \phi_{i,s}(1 - D_i)\right] , \qquad D_i = 1 - \left(\frac{f_{i,s}}{f_{i,\text{tot}}}\right)^2 \frac{m_s}{m_{\text{tot}}} ,$$

where $\phi_{i,\text{tot}}$ is the loss angle of coated disk and $\phi_{i,s}$ is the loss angle of the substrate. D_i is

Content from this work may be used under the terms of the Creative Commons Attribution 3.0 licence. Any further distribution of this work must maintain attribution to the author(s) and the title of the work, journal citation and DOI.
Published under licence by IOP Publishing Ltd

IOP Conf. Series: Journal of Physics: Conf. Series **957** (2018) 012006 doi:10.1088/1742-6596/957/1/012006

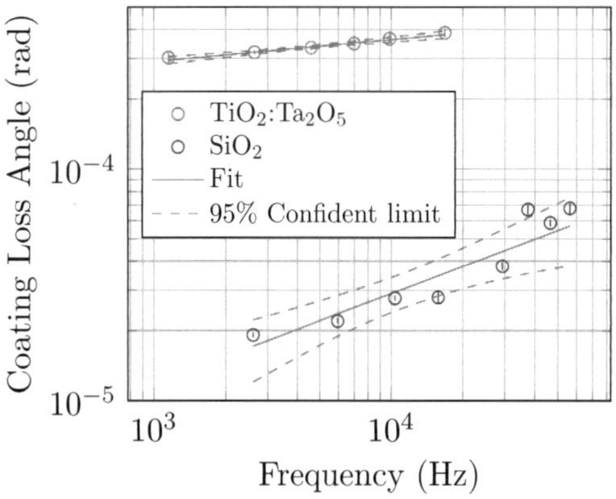

	$TiO_2:Ta_2O_5$	SiO_2
a (10^{-4})	1.5 ± 0.3	3 ± 1
b	0.11 ± 0.04	0.25 ± 0.05
n@1064nm	2.07 ± 0.02	1.45 ± 0.01
ρ $[g/cm^3]$	6.65 ± 0.06	2.15 ± 0.06

Figure 1: (Color online) Coating loss, fit parameters and optical properties of standard materials deposited by GC and annealed at 500°C for 10 hours.

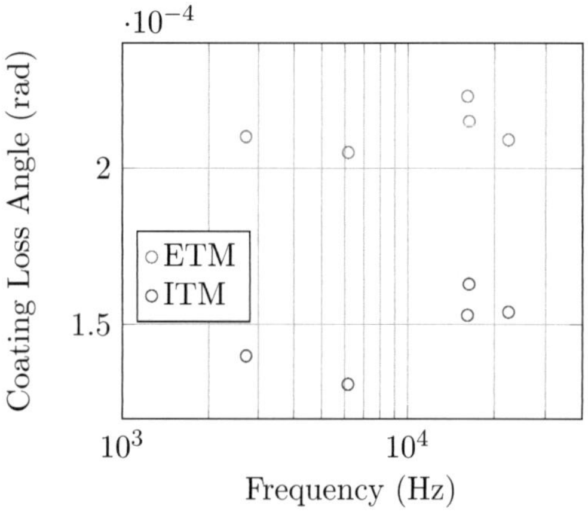

ETM	
n° of doublets	18
ratio $TiO_2:Ta_2O_5/SiO_2$	0.56
transmission	1 ppm
ITM	
n° of doublets	8
ratio $TiO_2:Ta_2O_5/SiO_2$	0.32
transmission	1.4%

Figure 2: (Color online) Coating loss and stack properties of ITM and ETM HR mirrors deposited by GC and annealed at 500°C for 10 hours [6].

the so-called *dilution factor* which can be related to $f_{i,s}$, $f_{i,\mathrm{tot}}$, m_s and m_{tot} [8], that are the frequencies and the mass of the sample before and after the coating deposition, respectively.

1. Standard materials in gravitational-wave interferometers

HR coatings of Advanced LIGO and Advanced Virgo are Bragg reflectors of alternate titania-doped tantala ($TiO_2:Ta_2O_5$) and silica (SiO_2) layers [1, 2]. Fig. 1 shows the mechanical loss of these materials, which seems to follow a power-law function $\phi_c = a \cdot f^b$. The loss angles of the HR coatings are shown on Fig. 2, together with their properties. The end mirror (ETM) coating has higher loss angle than the input mirror (ITM) coating because of its higher ratio $TiO_2:Ta_2O_5/SiO_2$.

2. Optimization

2.1. Doping

The purpose of TiO_2 doping is to increase Ta_2O_5 refractive index and reduce its loss angle. Increasing the refractive index contrast in the HR stack would allow to decrease the HR coating thickness, at constant reflectivity. Fig. 3a shows Ta_2O_5 coating loss as function of doping. The

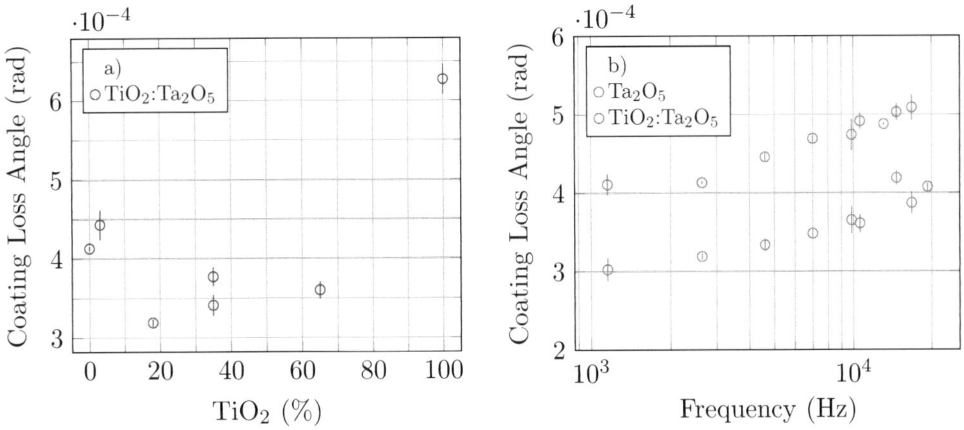

Figure 3: (Color online) a) Coating loss of TiO_2:Ta_2O_5 as function of TiO_2 doping. b) Comparison of Ta_2O_5 and 18%-doped TiO_2:Ta_2O_5 coating loss. All GC samples annealed at 500℃ for 10 hours (100% TiO_2 coating is crystallized).

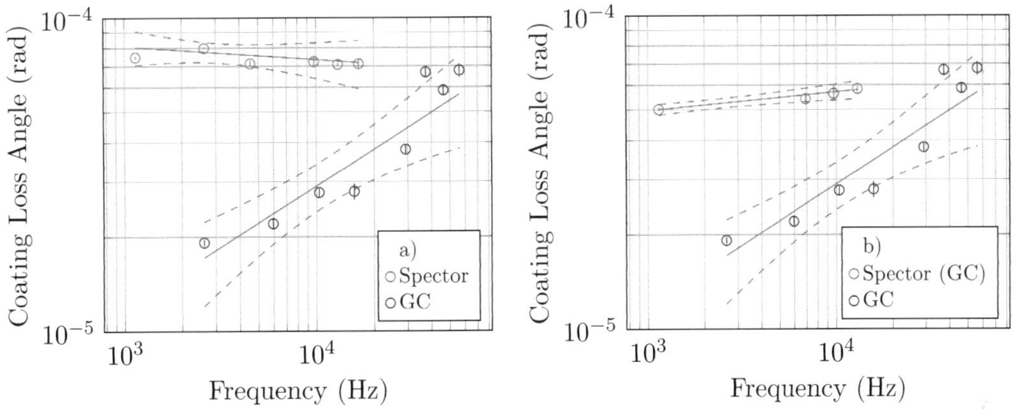

Figure 4: (Color online) Coating loss of SiO_2 deposited in the GC and in the Spector: a) different deposition parameters. b) same GC deposition parameters. All samples annealed at 500℃ for 10 hours. Fit model is the same as in Fig. 1.

current doping value in GW detectors is 18%, which yelds a minimum loss but a refractive index only slightly higher than that of Ta_2O_5. As shown by Fig. 3b, 18%-doped $\phi_{TiO_2:Ta_2O_5}$ is lower than $\phi_{Ta_2O_5}$ by $\sim$25%. Increasing TiO_2 concentration will increase TiO_2:Ta_2O_5 refractive index, while $\phi_{TiO_2:Ta_2O_5}$ for $TiO_2 \geq$40% can not be predicted and needs further investigation.

2.2. Deposition parameters

Fig. 4a shows coating loss of SiO_2 deposited by GC and Spector using their respective standard deposition parameters. It is clear that by using different parameters the same material gets different properties: GC parameters yield lower coating loss. For further test, SiO_2 has been deposited in the Spector with GC parameters. As Fig. 4b shows, Spector coating loss is lower but still higher than GC coating loss, because of the different configuration of the coaters. Coating losses of Ta_2O_5 deposited using different coaters have different values before annealing but converge toward a common limit value after, as shown in Fig. 7, suggesting that annealing

	SiO_2		Ta_2O_5		
	Spector	Spector as GC	GC	Spector	DIBS
a (10^{-4})	1.1 ± 0.2	0.33 ± 0.01	1.89 ± 0.09	2.6 ± 0.5	2.08 ± 0.08
b (10^{-1})	-0.4 ± 0.3	0.59 ± 0.05	1.00 ± 0.05	0.67 ± 0.08	0.88 ± 0.04
n	1.474 ± 0.005	1.468 ± 0.005	2.03 ± 0.02	2.08 ± 0.02	2.014 ± 0.005
ρ $[g/cm^3]$	2.34 ± 0.01	2.33 ± 0.05	7.34 ± 0.07	7.5 ± 0.1	6.9 ± 0.2

Table 1: Coating loss, fit parameters ($\phi_c = a \cdot f^b$), refractive index n at $\lambda = 1064$ nm and density ρ of materials deposited in different coaters.

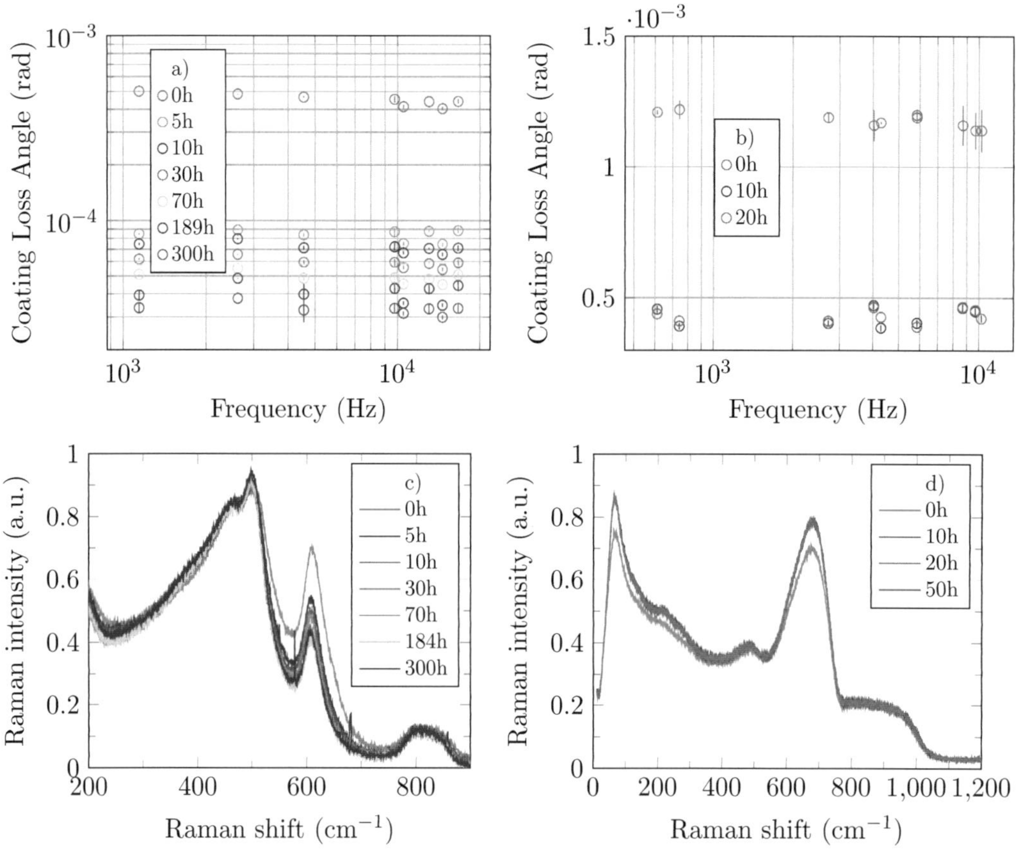

Figure 5: (Color online) Effect of annealing duration Δt [9]. top row: a) loss of SiO_2, b) loss of Ta_2O_5; bottom row: c) structure of SiO_2, d) structure of Ta_2O_5.

'deletes' the deposition history of the sample. Material properties are listed in table 1.

2.3. Post-deposition annealing

Annealing parameters are of fundamental importance for the purpose of reducing coating thermal noise. The problem is to find the optimal annealing temperature T_a and duration Δt, avoiding coating crystallization which would increase optical loss by scattering and absorption. In Fig. 5 is shown the effect of increasing Δt with T_a =500°C constant. SiO_2 loss decreases and this behaviour has a structural counterpart. SiO_2 is composed of tetrahedral units arranged in

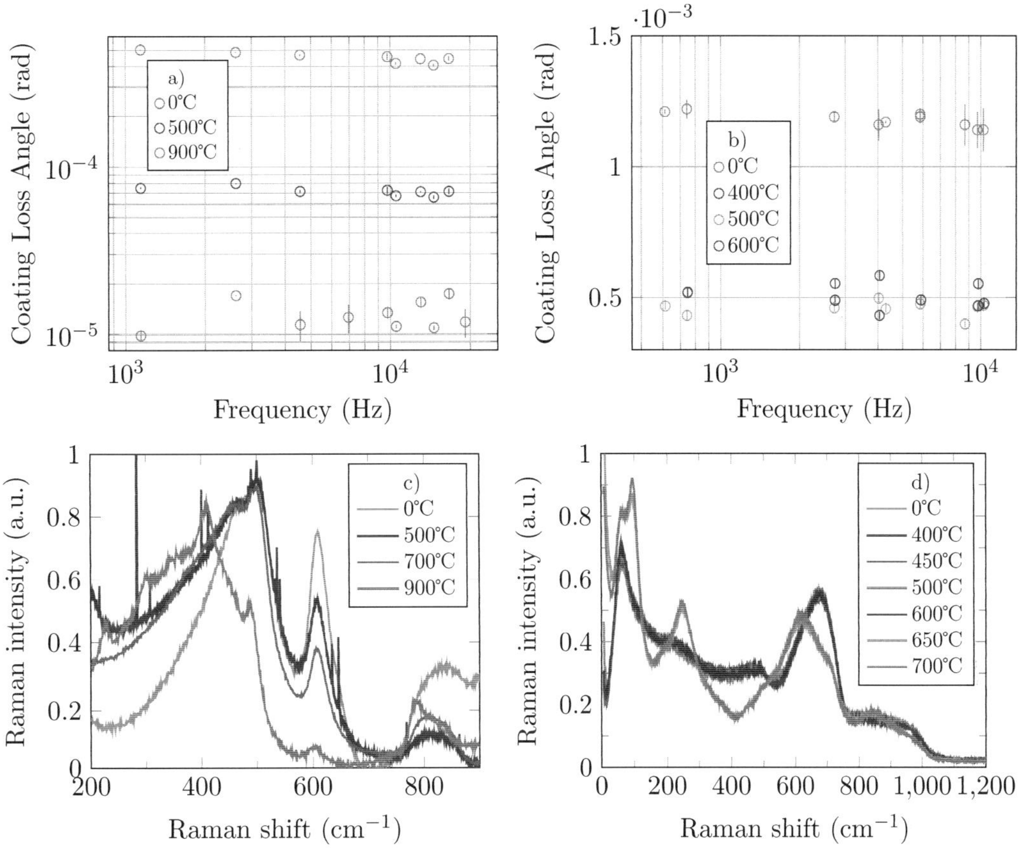

Figure 6: (Color online) Effect of annealing temperature T_a. top row: a) loss of SiO_2, b) loss of Ta_2O_5; bottom row: c) structure of SiO_2, d) structure of Ta_2O_5.

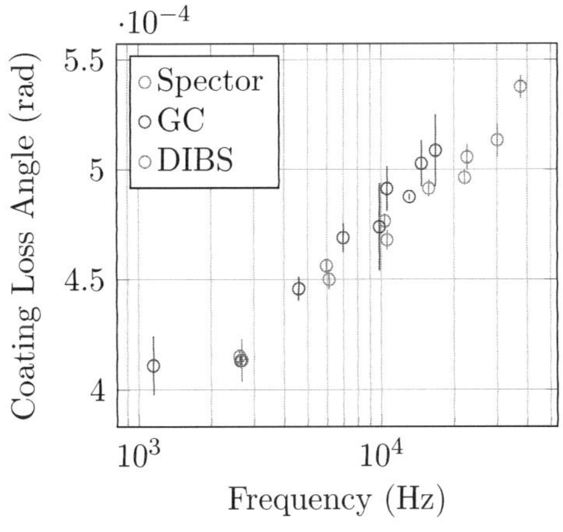

Figure 7: (Color online) Loss of Ta_2O_5 coatings deposited by different coaters and annealed at 500°C for 10h.

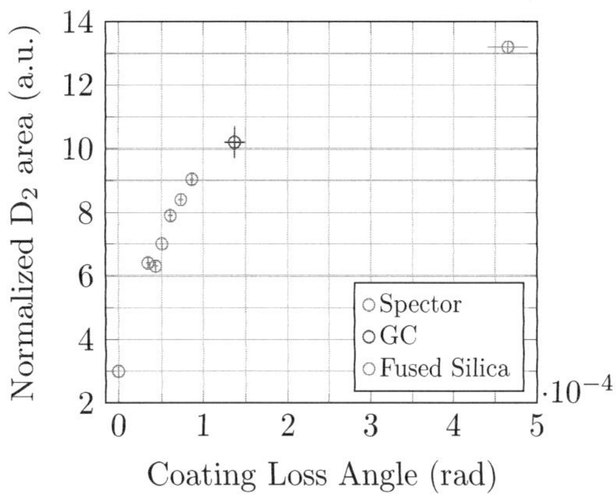

Figure 8: (Color online) Correlation between D_2 area and loss in SiO_2.

rings of different size [10] and the area of the D_2 band near 600 cm^{-1} is associated to 3-fold ring population [11]. A correlation between coating loss and D_2 has been found, suggesting that SiO_2 loss increases with the 3-fold ring population [9]. This correlation holds for different kinds of SiO_2, coating and bulk (Fig. 8). On the other hand, Ta_2O_5 loss does not change for $\Delta t \geq$10h and its structure evolves only for $\Delta t \leq$10h. Fig. 6 shows coating loss and structure for increasing T_a, with Δt =10h constant. SiO_2 loss decreases and its structure evolves considerably. Surprisingly, crystallization occurs at T_a =1000°C. For Ta_2O_5, coating loss is roughly constant for T_a >500°C and its structure does not change up to T_a =600°C, when crystallization occurs.

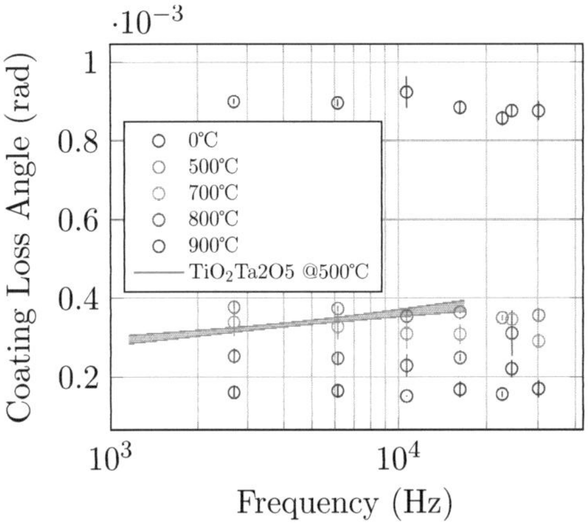

Figure 9: (Color online) Si_3N_4 loss for different T_a, compared to loss of TiO_2:Ta_2O_5 annealed at T_a =500°C.

3. New material

TiO_2:Ta_2O_5 could be replaced by a material with lower mechanical loss and possibly higher refractive index. Here silicon nitride (Si_3N_4) is proposed, which features high refractive index [12] and very low mechanical loss [13]. Usually, Si_3N_4 is deposited by low pressure chemical vapour deposition (LPCVD). However, LPCVD Si_3N_4 might suffer from hydrogen contamination and thickness uniformity issues, which are not compatible with the stringent optical specifications required for GW detectors. Instead, IBS Si_3N_4 can be developed in the GC for deposition on large optics. Fig. 9 shows a comparison between TiO_2:Ta_2O_5 and IBS Si_3N_4, this latter being annealed at different temperatures. Si_3N_4 loss decreases significantly at T_a =900°C. Thus, one could increase the annealing temperature of the entire HR stack to decrease also SiO_2 loss, eventually reducing the coating loss of the whole HR stack.

4. Conclusions

Coating materials of all present GW-detectors have been extensively characterized, showing a frequency-dependent loss angle. These standard materials can be optimized in different ways. The first approach is to increase the TiO_2 content in TiO_2:Ta_2O_5. Another option is to work on deposition parameters, in order to tune the optical and mechanical properties of the materials. In particular, the current GC configuration seems particularly well suited to deposit low loss SiO_2. In the case of Ta_2O_5, the effect of different deposition parameters is erased by the first few hours of post-deposition annealing. For 400°C< T_a <600°C and 10h< Δt <50h Ta_2O_5 loss shows null or limited evolution and its structure is frozen in a stable configuration. In the case of SiO_2, the annealing parameters T_a and Δt have a significant impact on mechanical loss and coating structure. A correlation is found between D_2 peak area in the Raman spectra, associated to the three fold ring population, and mechanical loss. IBS Si_3N_4 is an interesting new possibility to replace TiO_2:Ta_2O_5 because of its low mechanical loss. Furthermore, Si_3N_4 can be annealed at higher temperature than TiO_2:Ta_2O_5, reducing also SiO_2 coating loss angle and thus the loss of the whole HR coating.

References

[1] Aasi J *et al.* 2015 *Classical and Quantum Gravity* **32** 074001.
[2] Acernese F *et al.* 2015 *Classical and Quantum Gravity* **32** 024001.

[3] Aso Y, Michimura Y, Somiya K, Ando M, Miyakawa O, Sekiguchi T, Tatsumi D and Yamamoto H (The KAGRA Collaboration) 2013 *Phys. Rev. D* **88**(4) 043007
[4] Prato M, Chincarini A, Gemme G and Canepa M 2011 *Thin Solid Films* **519** 2877 – 2880 ISSN 0040-6090 5th International Conference on Spectroscopic Ellipsometry (ICSE-V)
[5] Cesarini E, Lorenzini M, Campagna E, Martelli F, Piergiovanni F, Vetrano F, Losurdo G and Cagnoli G 2009 *Rev. Sci. Instrum.* **80** 053904.
[6] Granata M *et al.* 2016 *Phys. Rev. D* **93**(1) 012007.
[7] Nowick A and Berry B 1972 *Anelastic relaxation in crystalline solids* (Academic Press)
[8] Li T *et al.* 2014 *Phys. Rev. D* **89**(9) 092004.
[9] Granata M *et al.* 2017 *arXiv preprint arXiv:1706.02928.*
[10] Jin W, Kalia R K, Vashishta P and Rino J P 1994 *Phys. Rev. B* **50**(1) 118–131.
[11] Pasquarello A and Car R 1998 *Phys. Rev. Lett.* **80**(23) 5145–5147.
[12] Chao S 2017 *LIGO-G1700304.*
[13] Liu X, Metcalf T H, Wang Q and Photiadis D M 2007 *MRS Proceedings* **989.**

IOP Conf. Series: Journal of Physics: Conf. Series **957** (2018) 012007 doi:10.1088/1742-6596/957/1/012007

Noise Budget and Interstellar Medium Mitigation Advances in the NANOGrav Pulsar Timing Array

T. Dolch[1] (for the NANOGrav Collaboration), S. Chatterjee[2], J. M. Cordes[2], P. B. Demorest[3], J. A. Ellis[4], M. L. Jones[4,5], M. T. Lam[4,5], T. J. W. Lazio[6], L. Levin[7], M. A. McLaughlin[4,5], N. T. Palliyaguru[8], D. R. Stinebring[9]

[1]Department of Physics, Hillsdale College, 33 E. College Street, Hillsdale, MI 49242, USA
[2]Astronomy Department, Cornell University, Ithaca, NY 14853, USA
[3]National Radio Astronomy Observatory, 1003 Lopezville Rd., Socorro, NM 87801, USA
[4]Department of Physics and Astronomy, West Virginia Univ., Morgantown, WV 26506, USA
[5]Center for Gravitational Waves and Cosmology, West Virginia University, Chestnut Ridge Research Building, Morgantown, WV 26505
[6]Jet Propulsion Laboratory, California Institute of Technology, 4800 Oak Grove Drive, Pasadena, CA 91106, USA
[7]Jodrell Bank Centre for Astrophysics, School of Physics and Astronomy, The University of Manchester, Manchester M13 9PL, UK
[8]Department of Physics, Texas Tech University, Box 41051, Lubbock, TX 79409, USA
[9]Department of Physics and Astronomy, Oberlin College, Oberlin, OH 44074, USA

E-mail: `tdolch@hillsdale.edu`

Abstract. Gravitational wave (GW) detection with pulsar timing arrays (PTAs) requires accurate noise characterization. The noise of our Galactic-scale GW detector has been systematically evaluated by the Noise Budget and Interstellar Medium Mitigation working groups within the North American Nanohertz Observatory for Gravitational Waves (NANOGrav) collaboration. Intrinsically, individual radio millisecond pulsars (MSPs) used by NANOGrav can have some degree of achromatic red spin noise, as well as white noise due to pulse phase jitter. Along any given line-of-sight, the ionized interstellar medium contributes chromatic noise through dispersion measure (DM) variations, interstellar scintillation, and scattering. These effects contain both red and white components. In the future, with wideband receivers, the effects of frequency-dependent DM will become important. Having anticipated and measured these diverse sources of detector noise, the NANOGrav PTA remains well-poised to detect low-frequency GWs.

1. Introduction

The NANOGrav PTA collaboration currently times 71 MSPs (as of 2018 January 1) in a long-term campaign with the Arecibo Observatory (AO) and the Green Bank Telescope (GBT). Each pulsar is timed roughly monthly, with a small number of high-precision pulsar timed weekly, by one or both telescopes for ~20 min per observation. The pulse profiles (intensities as a function of pulse phase, radio frequency, and time during the observation) are "folded", or time-averaged over a number of pulse periods, resulting in high S/N times-of-arrival (TOAs) when compared with a local high-precision hydrogen maser time standard. Within NANOGrav, the various working groups function together to maintain long-term pulsar timing and produce its data products. Outside NANOGrav, the collaboration's efforts are joined with the European Pulsar

Content from this work may be used under the terms of the Creative Commons Attribution 3.0 licence. Any further distribution of this work must maintain attribution to the author(s) and the title of the work, journal citation and DOI.
Published under licence by IOP Publishing Ltd

IOP Conf. Series: Journal of Physics: Conf. Series **957** (2018) 012007 doi:10.1088/1742-6596/957/1/012007

Timing Array (EPTA) collaboration and the Parkes Pulsar Timing Array (PPTA) collaboration as the International Pulsar Timing Array (IPTA).

As a National Science Foundation Physics Frontiers Center, the NANOGrav collaboration produces periodic public data releases. Each release (including associated metadata) is based on the deterministic timing model that can be updated as a result of the NANOGrav observations. The residuals between the observed and predicted TOAs (utilizing the uncertainties in timing residuals, also known as "RMS residuals") would become significantly non-zero in the presence of unmodeled physical processes. As for the signature of GWs in pulsar timing residuals, their significance in an individual pulsar may be low. However, GWs with light-year wavelengths passing through solar system should perturb TOAs from pulsars, each with a different line-of-sight (LoS), in a correlated manner [1]. The residuals from the NANOGrav dataset, begun in 2005, are sensitive to GWs in the nHz regime of the GW spectrum. The most likely GW source expected for PTA detection is a stochastic background of GWs from merging supermassive black hole binaries (SMBHBs), although continuous wave (CW) sources from individual SMBHBs in pre-merger orbits are also a possibility, as are memory bursts from the instants of SMBHB merger. In addition to SMBHBs, cosmic strings are also a potential GW source.

Each pair of pulsars in the PTA, in principle up to $\binom{71}{2}$ pairs, is analagous to the two arms of a laser interferometer GW detector. Just as a ground-based detector has its own noise spectrum as a function of GW frequency, so does every pulsar have its noise spectrum. Because a PTA is a Galaxy-sized GW detector, noise and signal must be measured together. Each NANOGrav data release now includes updated noise models, jointly derived by fitting all the pulsars' data from 2005 to the present. Some parameters correspond to individual pulsars, and others correspond to the Earth's observing conditions (for example, the solar system ephemeris model, which determines the barycentric position of the observatory when a particular observation occurs). Fitting for all-PTA noise models has been a primary activity of the Detection and Timing working groups within NANOGrav.

The Noise Budget (NB) working group provides an additional approach to noise characterization, in which each pulsar's noise spectrum is understood as a sum of contributions from various physical process appropriate to that pulsar. Just as all-PTA noise fits create a noise spectrum from the 71 pulsars jointly, individual noise measurements characterize a noise spectrum for a particular pulsar. This second measurement method approaches the ideal of an in-situ measurement of the noise in each detector component. In particular, noise processes can be characterized as either chromatic or achromatic. Growing achromatic noise in multiple pulsars, for example, would be consistent with a stochastic GW background. Chromatic noise, however, is consistent with unidentified processes affecting pulsar signal propagation in the ionized interstellar medium (IISM), which are radio-frequency dependent, and therefore more easily mitigated. Additional, the NB group can identify long-term observing strategies that can minimize noise.

Measuring and ultimately mitigating noise due to the IISM is under the purview of NANOGrav's Interstellar Medium Mitigation (IMM) group. The primary way that the IISM affects radio propagation is through dispersive delays, in which for a particular pulsar, TOA $\propto 1/\nu^2$, where ν is the radio frequency of received pulsar emission. The delay is proportional to DM, which is a measurement of the number of electrons along the LoS. The effects of dispersion are removed from the data in real time with the coherent dedispersion algorithm in pulsar backends; however, this removal assumes a particular DM value. DM for each pulsar varies by one part in 1000 of a fiducial DM in a timing model, enough to affect TOAs on timescales of μs or greater. NANOGrav observations are carried out at 2 or 3 center frequencies per pulsar, so that DM can be measured at every epoch. In addition to DM variations, diffractive interstellar scintillation (DISS) through a single thin phase-screen IISM overdensity perturbs TOAs with a $1/\nu^{4.4}$ dependence for an IISM distribution obeying a Kolmogorov turbulence distribution.

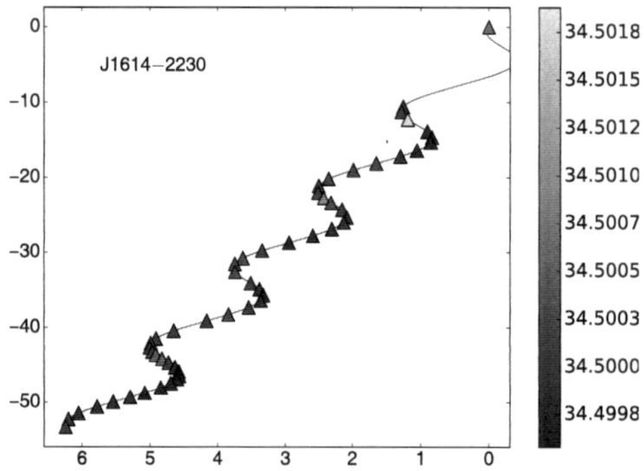

Figure 1. from [3]: the trajectory of PSR J1614–2230 is plotted with color mapping of the DM at each epoch. The axes depict the space traversed at the phase screen in au in the R.A. and decl. directions. The pulsar's motion starts at (0, 0). Pulsars closer to the ecliptic will show a tighter sinusoid than those farther away. Such trajectory plots can be used to show limited localized structure. Note the high DM regions in the same part of its trajectory every year, suggesting a localized overdensity along the line of sight to the pulsar.

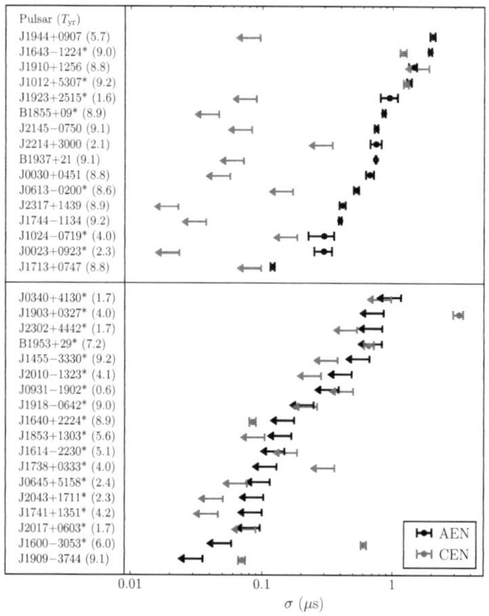

Figure 2. from [4]: Measured achromatic (AEN, black) and chromatic EN (CEN, red) for each NANOGrav pulsar ranked by AEN. The numbers in parentheses are the total years observed. Asterisks indicate pulsars for which the white-noise model is an upper limit for at least one receiver. Detections of AEN (top) are separated from the limits (bottom).

2. Advances in Interstellar Medium Mitigation

The IMM working group has analyzed potential effects of the IISM that would not ordinarily be characterized in the standard process of TOA and timing residual generation. Timing models are inherently deterministic. Variations in DM are not generally predictable, and it is a standard PTA practice to measure the DM every epoch utilizing TOAs at widely-spaced frequencies. DM variations do, however, have both deterministic and stochastic components, the predictable component arising from the deterministic LoS change due to the Earth's and a pulsar's velocities; see the IMM group's results in [3] (Figure 1; which is the example of PSR J1614–2230) and [5]. Multipath diffractive scattering also occurs in addition to DM delays, when an overdensity in the IISM temporarily redirects the signal's propagation paths. The diffraction creates a changing interference pattern across observing bandwidth, and also forward scatters a pulse's flux in time with scattering timescale τ_{d}. Scattering timescales are nontrivial to measure. A constant τ_{d} would be irrelevant for GW detection. If significantly varying, however, the scattering tails of received pulses would provide another source of stochastic noise in timing residuals [6]. In principle, re-measuring a pulsar's scattering tail at every epoch is possible, but a current implementation is not computationally trivial across the entire backend receiver bandwidth (at AO, the Puerto Rican Ultimate Pulsar Processing Instrument, PUPPI; at the GBT, the Green Bank Ultimate Pulsar Processing Instrument, GUPPI; [7]).

In the NANOGrav nine-year data release [8], the (then) 37 pulsars were shown to generally not have significant variations in scattering parameters [9], compared to the RMS residuals from all noise processes. Scattering timescales at each observing epoch were computed not from directly measuring scattering tails, but by utilizing dynamic spectra (DS; signal intensity as a function of both time and radio frequency). The interference pattern across bandwidth due to the multipath scattering of the on-pulse signal at any given time is directly related to τ_d according to $\tau_d \sim 1/(2\pi\Delta\nu_d)$, where $\Delta\nu_d$ is the diffractive bandwidth. The constructive interference peaks (or "scintles" with a characteristic width equal to $\Delta\nu_d$) change on diffractive timescales. The unpredictable filling pattern of scintles in a DS is a source of white noise in timing residuals, $\sigma_{\rm DISS}$. Scattering timescale characterization using dynamic spectra (as opposed to IISM characterization through TOAs alone) is a good example of using ancillary data in PTA science with the aim of GW detection (see also [10]).

The low level of scattering variation was to some degree a selection effect, because a pulsar with highly varying τ_d values would not have originally been selected for inclusion in a PTA. In future data releases, however, sensitivity to GWs will have increased, such that the τ_d measurements in [9] may need to be corrected through techniques such as cyclic spectroscopy ([11], [12], [13], [22]), which is another project within the IMM group.

Another possible unquantified stochastic noise process could arise from more complex scattering processes detectable with very wide-bandwidth receivers. The analysis in [2] shows that highly-scattered pulsar signals may, along different ray paths, propagate through significantly different numbers of electrons, integrated along their respective paths, before their refocusing. As a result, over very wide bandwidths, pulsar signals might follow a different law than TOA $\propto 1/\nu^2$ (because otherwise, the standard cold-plasma propagation law assumes an essentially 1D propagation). This prediction is not only significant for future wideband receivers, but also an interesting explanation for the excess chromatic noise found in many NANOGrav pulsars (to be described in the following section). In the future, a more complex frequency-dependent propagation law might be a standard component of timing models.

3. Advances in Noise Budget Descriptions

The NB working group has constructed bottom-up noise models for each individual pulsar. It is essential to have such an approach along with top-down, all-PTA global noise modeling; a stochastic background of GWs would initially show up as a source of red noise in the PTA as a whole. However, individual-pulsar noise modeling, advancing significantly with each data release, becomes increasingly useful as both a prior on all-PTA noise modeling, as well as an independent pipeline to characterize the accumulation of red noise in pulsars.

Red noise can be characterized [14] as having chromatic and achromatic components. Stochastic noise processes intrinsic to a pulsar such as spin noise, intrinsic to the observation position such as noise from ephemeris model uncertainties, or even extrinsic "red noise" from a GW background (more properly, a signal with a red power spectrum) have no radio frequency dependence. Stochastic noise processes arising from the pulsar signal's interaction with the IISM should have a radio frequency dependence. In a recent study from the NB group [4], we conducted measurements of red noise in each NANOGrav pulsar, separating the noise out into chromatic and achromatic components.

The red noise measurements in [4] (Figure 2) are also referred to as "excess noise", in other words, noise in excess of well-characterized white noise processes. The white noise models come from [15] (Figure 3), in which bottom-up white noise measurements are conducted on NANOGrav pulsars. The global PTA fit yields three white noise components to TOAs from the TEMPO2 software package: "EFAC", which simply multiplies each TOA uncertainty by a scale factor; "EQUAD", which adds a white noise component in quadrature to each TOA uncertainty (and this performed independently at each receiver frequency) and "ECORR", which adds the same white noise component to each receiver frequency, in other words, as achromatic noise.

In the top-down approach, significant ECORR values can be likely identified with a well-known source of achromatic white noise in individual pulsars: pulse phase jitter. However, the white noise modeling in [15] can use the residuals from all epochs to directly measure the white noise due to pulse phase jitter in each pulsar, by fitting for the white noise in the RMS residual values that does not vanish even in the highest S/N residuals. This method was first applied in [19] and then extended to all NANOGrav pulsars in [20]. Additionally, [15] utilizes the scattering measurements in [9] (deriving from DS not utilized in a global PTA fit) to characterize $\sigma_{\rm DISS}$ for each pulsar, separating $\sigma_{\rm DISS}$ out from pulse phase jitter explicitly.

Having characterized the white noise for each pulsar individually, [4] then computes the excess red-noise not described by the white noise models (Figure 3). This results in either noise upper limits or significant measurements depending on the pulsar. We find that 16 of the 37 nine-year data release pulsars are found to have achromatic excess noise at the level of 100 ns or greater. We also find that 7 pulsars have significant chromatic excess noise. Taking the chromatic measurements and upper limits together, and plotting each pulsar's noise value (or limit) with respect to its DM, an approximate proportional relationship emerges, albeit with outliers. The unmodeled chromatic noise may be due to either $\tau_{\rm d}$ variation (as in [9]) and/or due to additional radio frequency-dependencies such as that proposed in [2].

As for the excess achromatic noise, there are a number of plausible noise mechanisms: low-level, unmitigated, broadband radio frequency interference (RFI); polarization calibration errors; and yet-umodeled wandering of planetary ephemeris parameters. An intriguing possible explanation for the growing red noise measurements in the nine-year data release (as compared to the five year data release [16]) is the presence of a GW background. However, a convincing eventual detection would also include *correlated* noise [1] between pulsars in addition to a common red signal. With additional NANOGrav data releases utilizing both AO and the GBT, the further dissection of achromatic noise into the possible components described here is highly likely.

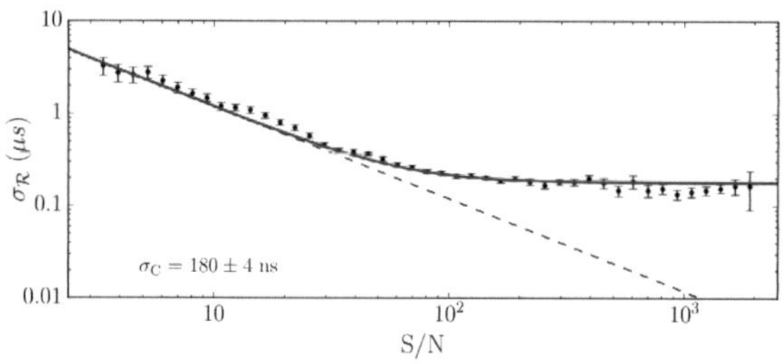

Figure 3. from [15]: analysis of residuals for PSR J1713+0747 observed at 1400 MHz with AO, containing the highest S/N residuals in our sample. RMS residuals (σ_R) are in bins of S/N. The dashed line is the predicted TOA fitting error (not a fit to the points on the graph) based on the pulse template shape, while the solid line shows the estimated RMS residual that includes a S/N-independent term.

In order to best make use of future NANOGrav observing time, another project of the NB working group [20] has been the determination of the optimal observing center frequencies and bandwidths for each pulsar, for each telescope. Currently, the standard timing data from the GUPPI and PUPPI backends is taken at up to three widely separated center frequencies. Future receivers are expected to span an octave of bandwidth or more. The study in [20] applies to a future wide-bandwidth receiver, suggesting its optimization for a specific PTA, given the varieties of frequency-dependent IISM noise in timing residuals present across the most significant NANOGrav pulsars.

Studying the NB of MSPs also results in ancillary science. The 24-hr global campaign [19] was a 9-telescope observation by the IPTA conducted for the purpose of studying PSR J1713+0747's noise budget on multiple-hour timescales. This particular MSP was especially important for noise characterization as the most sensitive individual pulsar observed by the entire IPTA. As a result of the campaign, a new GW limit was obtained [21] in the GW frequency range 10^{-5} – 10^{-3} Hz, both for the entire sky and for any GW sources in the direction of the pulsar.

4. Conclusions

The efforts of the NB and IMM working groups have characterized the noise, largely due to IISM effects, of the pulsars in the NANOGrav PTA as a whole. These results include models of the IISM and its interaction with pulsar signals ([5], [3]). In particular, the models of [3] are applicable to future datasets, for which the deterministic and stochastic components of DM variations can be more confidently separated. Direct white noise measurements of [15] and [9] enable a high-confidence identification of intraday noise sources with specific physical processes, leading in turn to a measurement of each pulsar's excess noise, which can be further decomposed into chromatic red noise and achromatic red noise. The achromatic red noise component may itself already contain timing perturbations due to a stochastic GW background, as NANOGrav approaches the sensitivity in which predicted background levels should be present [17]. However, more definite statements will require discriminating GW signatures from those of other red noise processes, highlighting the importance of continual noise budgeting. The value of continuing NB and IMM characterization is not only for GW detection; indeed, the ancillary ISM science can reveal unusual events present in DM(t) data, such as that reported by the PPTA in PSR J1713+0747 [10]. Such events inform plasma lensing models which are also relevant to studying propagation effects in fast radio bursts [18].

5. Acknowledgments

The NANOGrav project receives support from National Science Foundation (NSF) Physics Frontier Center award number 1430284 and NSF-PIRE program award number 0968296. TD acknowledges travel support from the NANOGrav PFC award. Part of this research was carried out at the Jet Propulsion Laboratory, California Institute of Technology, under a contract with NASA. JAE acknowledges support by NASA through Einstein Fellowship grant PF4-150120.

6. References

[1] Hellings, R. W., & Downs, G. S. 1983, *Astrophysical Journal* **265** L39
[2] Cordes, J. M., Shannon, R. M., & Stinebring, D. R. 2016, *Astrophysical Journal* **817** 16
[3] Jones, M. L., McLaughlin, M. A., Lam, M. T., et al. 2017, *Astrophysical Journal* **841** 125
[4] Lam, M. T., Cordes, J. M., Chatterjee, S., et al. 2017, *Astrophysical Journal* **834** 35
[5] Lam, M. T., Cordes, J. M., Chatterjee, S., et al. 2016, *Astrophysical Journal* **821** 66
[6] Stinebring, D. 2013, *Classical and Quantum Gravity* **30** 224006
[7] DuPlain, R., Ransom, S., Demorest, P., et al. 2008, *SPIE Conference Series* **7019** 70191D
[8] The NANOGrav Collaboration, Arzoumanian, Z., Brazier, A., et al. 2015, *Astrophysical Journal* **813** 65
[9] Levin, L., McLaughlin, M. A., Jones, G., et al. 2016, *Astrophysical Journal* **818** 166
[10] Keith, M. J., Coles, W., Shannon, R. M., et al. 2013, *Monthly Notices of the Royal Astronomical Society* **429** 2161
[11] Palliyaguru, N., Stinebring, D., McLaughlin, M., Demorest, P., & Jones, G. 2015, *Astrophysical Journal* **815** 89
[12] Demorest, P. B. 2011, *Monthly Notices of the Royal Astronomical Society* **416** 2821
[13] Walker, M. A., Demorest, P. B., & van Straten, W. 2013, *Astrophysical Journal* **779** 99
[14] Cordes, J. M., & Shannon, R. M. 2010, (*Preprint* astro-ph/1010.3785)
[15] Lam, M. T., Cordes, J. M., Chatterjee, S., et al. 2016, *Astrophysical Journal* **819** 155
[16] Demorest, P. B., Ferdman, R. D., Gonzalez, M. E., et al. 2013, *Astrophysical Journal* **762** 94
[17] Chen, S., Middleton, H., Sesana, A., Del Pozzo, W., & Vecchio, A. 2017, *Monthly Notices of the Royal Astronomical Society* **468** 404
[18] Cordes, J. M., Wasserman, I., Hessels, J. W. T., et al. 2017, *Astrophysical Journal* **842** 35
[19] Dolch, T., Lam, M. T., Cordes, J., et al. 2014, *Astrophysical Journal* **794** 21
[20] Lam, M. T., McLaughlin, M. A., Cordes, J. M., Chatterjee, S., & Lazio, T. J. W. 2017, (*Preprint* astro-ph/1710.02272)
[21] Dolch, T., NANOGrav Collaboration, Ellis, J. A., et al. 2016, *Journal of Physics Conference Series* **716** 012014
[22] Archibald, A. M., Kondratiev, V. I., Hessels, J. W. T., & Stinebring, D. R. 2014, *Astrophysical Journal Letters* **790** L22

Structured Doctoral Education in Hannover - Joint Programme IMPRS-GW and *geo*-Q RTG

Fumiko Kawazoe[1] **and Sandra Bruns**[2]

[1]Institut für Gravitationsphysik der Leibniz Universität Hannover, Callinstr. 38, D-30167 Hannover, Germany

[2]Max-Planck-Institut für Gravitationsphysik (Albert-Einstein-Institut) Callinstr. 38, D-30167 Hannover, Germany

E-mail: fumiko.kawazoe@aei.mpg.de

Abstract. Two structured doctoral programmes that we have in Hannover, the IMPRS on Gravitational Wave Astronomy and SFB on relativistic geodesy and gravimetry with quantum sensors *geo*-Q, have not only become major resources for education in each field but have also started to provide substantial synergy to members of both programmes. Our strong cross-disciplinary approach to create a joint programme has received excellent feedback not only from researchers inside the programmes but also from various external committees. Building on experience that we have acquired over the last decade, we propose to set up a common doctoral programme within the international gravitational wave astronomy and physics. We envisage that with a common doctoral programme we will create a strong team of young researchers who will carry on building a strong network of third generation gravitational wave detectors and observatories.

1. Introduction

After the ground-breaking discovery of gravitational waves in 2015 [1], the overwhelming success of the LISA Pathfinder mission in 2016 [2], and the selection of the LISA mission [3] as the L3 mission in the future science programme of the European Space Agency ESA gravitational wave astronomy has more than ever a brighter and a louder future. Building space missions like LISA and third generation detectors like Einstein Telescope [4] is a long term undertaking which will be carried out by the students that we educate now. In order to bring those projects into success they need a high-quality education and a good team spirit with shared values. In Hannover we have two structured doctoral programmes that could be a model for building a strong network of the future missions and third generation detectors. One is the International Max Planck Research School on Gravitational Wave Astronomy (IMPRS-GW) and the other is a new addition called the *geo*-Q Research Training Group (RTG) for Earth scientists and physics. Recently we have successfully created a synergy between them by forming a joint doctoral programme. The merging of the two different programmes can serve as a role model for the entire field of gravitational wave research collaborators.

2. IMPRS-GW

IMPRS-GW is the single largest and most influential educational institution worldwide in the field of gravitational wave research. It has been a joint doctoral programme of the

Content from this work may be used under the terms of the Creative Commons Attribution 3.0 licence. Any further distribution of this work must maintain attribution to the author(s) and the title of the work, journal citation and DOI.
Published under licence by IOP Publishing Ltd

Albert Einstein Institute (AEI)(Divisions Gravitational Wave Astronomy, and Observational Relativity & Cosmology, and the Astrophysical and Cosmological Relativity division), the Leibniz Universität Hannover (LUH) (Institute for Gravitational Physics (IGP), Institute for Quantum Optics (IQO), and Institute for Theoretical Physics), and the Laser Zentrum Hannover (Division Laser Development), and the University of Potsdam. The programme combines the different fields of gravitational wave astronomy: from laser development, interferometry and quantum optics to data analysis, astrophysics, theory, and source modelling.

2.1. Scope and curriculum

The IMPRS-GW curriculum ensures a high-quality and comprehensive graduate education. It is our goal that theorists acquire a basic understanding of the experimental foundations of laboratory work in general and of interferometry, laser physics and quantum optics in particular. And experimentalists, on the other hand, are expected to acquire a basic understanding of data analysis, relativistic astrophysics and numerical relativity. We also aim to provide a system of shared values and principles concerning scientific behaviour and methodology. In order to achieve these goals we have altered our curriculum several times over the past decade and now found an optimum compromise between the traditional way of learning on-the-job by working as independent researchers alongside experienced colleagues and helpful peers on the one hand, and the best elements brought by the structured doctoral programme such as attending tailored lectures, group experiences and transferable skills education on the other hand. These transferable skills courses include intensive training from world's top-class trainers on "Scientific Writing", "Scientific Presenting", "Project Management", and " Good Scientific Practice". All of our students receive intensive advising through their own Thesis Advisory Committee consisting of at least three supervisors who they choose when they enrol at the university to study at our institute. We also provide ample opportunity to all our doctoral students to spend several months at an international collaborating research institution and all our doctoral students make presentations at one international conference and many German meetings and present their thesis work. The curriculum of the IMPRS-GW is based on three pillars of our education. They are the specialised lectures offered regularly at LUH, specialised lectures by AEI scientists which are video-transmitted to the other part of the institute, and our so-called IMPRS Lecture Weeks, introduced in 2008.

2.2. IMPRS-GW Lecture Weeks

These are weeks of block lectures in general relativity, theoretical astrophysics, data analysis & statistics, numerical relativity (alternating), and experimental gravitational wave physics (with changing topics), taking place at "isolated" locations in the surroundings of Europe. We have three lecture weeks per year. With this concept, we want to ensure that experimentalists also get some basic knowledge in the theoretical fields of gravitational wave astronomy and physics and theorists learn the basics of the experimental fields. The second important goal of these weeks is to bring the students of the different research fields and both institute locations together. Three lectures are given every day with a good balance of theoretical topics and experimental topics in the first two lecture weeks. In the afternoon and in the evening students are given plenty of time to deepen their understanding by asking questions, to present their thesis work, and to get to know each other and building network. In the third week a one-week project by student teams composed of theorists and experimentalists as well as advanced students and beginners is added. This project work is increasing considerably the active student participation through hands-on experience. Details of topics given at three lecture weeks are shown in figure 1.

	Lecture Week 1	Lecture Week 2	Lecture Week 3
General Relativity/ Astrophysics	- Special Relativity - Curved Coordinates, Equivalence Principle - Tensors and physics in a curved spacetime - Einstein equations, initial-value formulation - Linearized GWs	- Linearized theory, action on detect - Generation of GWs in linearized theory - Black holes I - Black holes II - Spherical stars	- Massive black holes in the Universe - Massive black hole dynamics - Evolution of single stars - Evolution of binary systems and stellar remnants
GW Interferome try	- GWs and their effect - Detection principle - Interferometer basics - Signal detection theory - Detector noise sources	- Power Spectral Density - Signal to Noise ratio - Adv. Ifo configurations - Control theory & practice - Gaussian optics	- Thermal Noise - Fluctuation-Dissipation Theorem - Non-classical light - Generation of squeezed light - Squ. light application & Standard Quantum Limit
Statistics/ Data Analysis	- Discrete differential operators - 3+1 split of spacetime - Different formulations of Einstein equations - Gauges, initial data and GW extraction - Introduction to relativistic hydro-dynamics	-Bayesian, Neyman- Pearson - Likelihood, matched filtering - Signal vetoes, chi-square, significance of signal - Gaussian and non-Gaussian noise and statistics - Multiple detector burst searches	Project Week

Figure 1. IMPRS-GW Lecture Week yearly curriculum. Three lectures are given every day, as a rule starting with a lecture with experimental topics, followed by a theory lecture, then another experimental lecture to keep a good balance of topics.

2.3. Outcome of the education
After 10 years of continuous effort in operating IMPRS-GW, 91 doctoral students have graduated and all of them have obtained PhD degrees. Assuming that the first set of doctoral students took four years to graduate, our alumni have been on the job market for the past six years. All of IMPRS-GW alumni have obtained a job as they graduated. As of today, more than half are working as postdocs all over the world, 25% have obtained a job in industry, a few alumni became professors in physics and others became teachers in high schools, as shown in figure 2. Our experience allows us to foresee that more alumni will shift into industry the longer they are on the market. Our approach to avoid turning graduate research and study into a type of rigid higher-level school system but to combine the best elements from all approaches have produced young researchers who are equipped with knowledge and skills that are highly sought after in both academic world and industry. This dual career nature of our alumni is a clear indicator that we have been successfully educating our doctoral students.

2.4. Evolving IMPRS-GW
Not only doctoral students but also postdocs and senior researchers from all over the world have joined our programme. For example in this Amaldi conference there were five scientists who participated in our lecture weeks as lecturers and we also found that one out of every five speakers of the conference have participated in our lecture weeks. This shows that we have become a major resource of education in the field of gravitational wave astronomy and physics. Recently we started to expand our programme by inviting people from outside of our

IOP Conf. Series: Journal of Physics: Conf. Series **957** (2018) 012008 doi:10.1088/1742-6596/957/1/012008

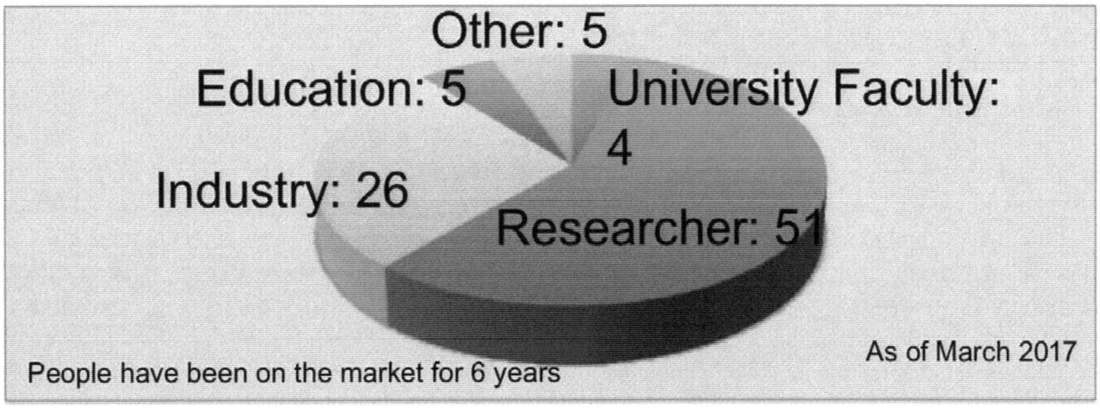

Figure 2. Type of jobs obtained by IMPRS-GW alumni. The result shows a dual career nature. We can foresee that more alumni will obtain jobs in industry the longer they are on the job market.

own institute to provide specialised experience that we do not have and to bring people of different groups together, e.g. "Quantum Optomechanics" by Markus Aspelmeyer (University of Vienna), Joint week with International Max Planck Partnership IMPP with University of Glasgow in 2015 and 2017, Earth scientists from University of Balearic Island, and lecturers from space industry.

2.5. Merging IMPRS-GW and geo-Q programme

IMPRS-GW is constantly evolving. Recently we put two graduate school programmes together. The other programme is the Sonderforschungsbereich (SFB)[1] *geo*-Q, which is supported by the German Research Foundation. The group consists of IGP, IQO and Institut of Geodesy (IfE) at the LUH, The National Metrology Institute of Germany (PTB) in Braunschweig near Hannover, and Center of Applied Space Technology and Microgravity (ZARM) in Bremen. *geo*-Q is centered at LUH, and there are some overlaps between the research topics and principal investigators (PIs) of the two programmes. Merging the two programmes allows us to therefore not only optimise learning process but also reduce teaching effort, since otherwise the PIs would have two programmes to support. In addition, by the merging method we obtain an increased level of interdisciplinarity which allows our students to acquire broadened and deepened understanding of their research field.

3. *geo*-Q

geo-Q aims to obtain better knowledge of the Earth gravity field by combining three pillars together. They are better accuracy of global gravity field using satellites with laser interferometer, measuring gravitational redshift to get very accurate gravitational potential with atomic clocks, and obtaining local gravity acceleration with quantum gravimeter. These are very different fields, but the structured doctoral programme of *geo*-Q, the *geo*-Q RTG, brings them together to form a strong team. All doctoral students of *geo*-Q are members of the *geo*-Q RTG, and they learn topics of relativity, theoretical geodesy, experimental geodesy, satellite designs, data analysis & statistics, and various experimental topics including optical clocks, matter wave interferometers, and laser interferometry. It is our goal that they will be the first generation of

[1] In English it is called the Special Research Area

relativistic geodesists who have understanding of all of these fields. The *geo*-Q RTG curriculum is based on that of the IMPRS-GW.

4. Joint programmes

At first we did not know to what level the two programmes could be merged but after a few trials we found that lecture weeks and transferable skills courses could be very well combined. Our joint lecture weeks host common lectures to students from both IMPRS-GW and the *geo*-Q RTG, such as basics of special relativity, basics of data analysis & statistics, and laser interferometry, and parallel sessions in which two separate lecturers are given for the two groups. In addition we have guest lectures for general education, and those are for students from both programmes. Such guest lectures include e.g. oceanography and astrophysics. Obtaining one additional seminar room enabled us to transform a normal lecture week into a joint lecture week. Students as well as lecturers have given excellent feedback on this joint programme and we are open to new suggestions of improvement when needed.

5. Conclusion and outlook

Not only have we created the joint programme between the IMPRS-GW and the *geo*-Q RTG, but also we aim to expand our activities to start a European wide structured doctoral programme in the future. Currently all our students involved in LISA and GEO600 [5] are our IMPRS-GW members. Our vision is to develop a world-wide educational programme in gravitational wave astronomy and physics, and to create a new generation of gravitational wave physicists who will carry on the projects of third generation detector network and future missions. As a first step we submitted a proposal for a joint European-wide doctoral programme with University of Lyon. We have already been actively exchanging students within the collaboration. However that is not enough, and we need a common curriculum, common advising, and possibly even a real joint doctoral degrees among our collaborating institutions. We strongly propose to let us build a world-wide programme for third generation detectors and future missions network together. It is straightforward to start based on an already functioning model like IMPRS-GW programme, which has successfully demonstrated a possible method for merging with a different programme such as *geo*-Q RTG. We envisage that it is relatively easy to start an European-wide joint lecture week. It is not complete nor perfect, but it is a start.

Acknowledgments

The authors would like to thank the DFG Sonderforschungsbereich (SFB) 1128 Relativistic Geodesy and Gravimetry with Quantum Sensors (geo-Q) for financial support for the *geo*-Q RTG, and the Max-Planck-Gesellschaft for granting the IMPRS-GW.

References

[1] Abbott B P *et al.* 2016 *Phys. Rev.* Lett **116** 061102
[2] Armano M *et al.* 2016 *Phys. Rev.* Lett **116** 231101
[3] Danzmann K *et al. Laser interferometer space antenna technical report* (LISA Consortium) URL `https://www.elisascience.org/files/publications/LISA_L3_20170120.pdf`
[4] Abernathy M *et al. ET-0106C-10 technical report*
[5] Dooley K L *et al.* 2016 *Class. Quantum Grav.* **33** 075009

IOP Conf. Series: Journal of Physics: Conf. Series **957** (2018) 012009 doi:10.1088/1742-6596/957/1/012009

Optical design and suspension system of the KAGRA output mode-cleaner

Junko Kasuya, John Winterflood[1], Ju Li[1] and Kentaro Somiya

Department of Physics, Tokyo Institute of Technology, Tokyo 152-8551, Japan
[1]Department of Physics, University of Western Australia, Perth 6009, Australia

E-mail: kasuya@gw.phys.titech.ac.jp

Abstract. KAGRA is a Japanese large scale, underground, cryogenic gravitational telescope which is under construction in the Kamioka mine. For using cryogenic test masses, the sensitivity of KAGRA is limited mainly by quantum noise. In order to reduce quantum noise, KAGRA employs an output mode-cleaner (OMC) at the output port that filters out junk light but allows the gravitational wave signal to go through. The requirement of the KAGRA OMC is even more challenging than other telescopes in the world since KAGRA plans to tune the signal readout phase so that the signal-to-noise ratio for our primary target source can be maximized. A proper selection of optical parameters and anti-vibration devices is required for the robust operation of the OMC. In this proceeding, we show our final results of modal-model simulations, in which we downselected the cavity length, the round-trip Gouy phase shift, the finesse, and the seismic isolation ratio for the suspended optics.

1. Introduction

Gravitational waves are ripples of space-time, first predicted by Albert Einstein nearly 100 years ago [1]. On September 14, 2015, two LIGO observatories succeeded in the first direct detection of gravitational waves and the first observation of a binary black hole merger [2]. At the dawn of the gravitational wave astronomy is approaching, it is essential for the gravitational wave detectors to collaborate together.

KAGRA is an interferometric gravitational telescope which is under construction in the Kamioka mine, Japan [3]. The interferometer has an 3-km Fabry-Perot cavity in each arm, a power-recycling cavity in the symmetric port, and a signal-recycling cavity in the anti-symmetric (AS) port (see Fig. 1, left). The two unique features of KAGRA are (i) the entire interferometer is built underground for less seismic vibration and (ii) the sapphire test masses are cooled down to a cryogenic temperature. The gravitational wave signal is obtained with a DC readout scheme [4], in which a fraction of the carrier light leaking to the AS port is used as a reference field to beat with the gravitational wave signals. To maximize the observation range to our primary target source: binary neutron stars, the signal-recycling cavity is detuned by 3.5 deg and the DC readout phase is tuned to 132 deg. The observation range is then 153 Mpc. In order to tune the readout phase, the reference field power may need to be as small as 1 mW, depending on the imbalance of the two arm cavities. Since the OMC needs to filter out and suppress other light coming out to the AS port with such a small amount of reference light, the requirement to the KAGRA OMC is even more challenging than the OMCs in other telescopes.

Content from this work may be used under the terms of the Creative Commons Attribution 3.0 licence. Any further distribution of this work must maintain attribution to the author(s) and the title of the work, journal citation and DOI.
Published under licence by IOP Publishing Ltd

IOP Conf. Series: Journal of Physics: Conf. Series **957** (2018) 012009 doi:10.1088/1742-6596/957/1/012009

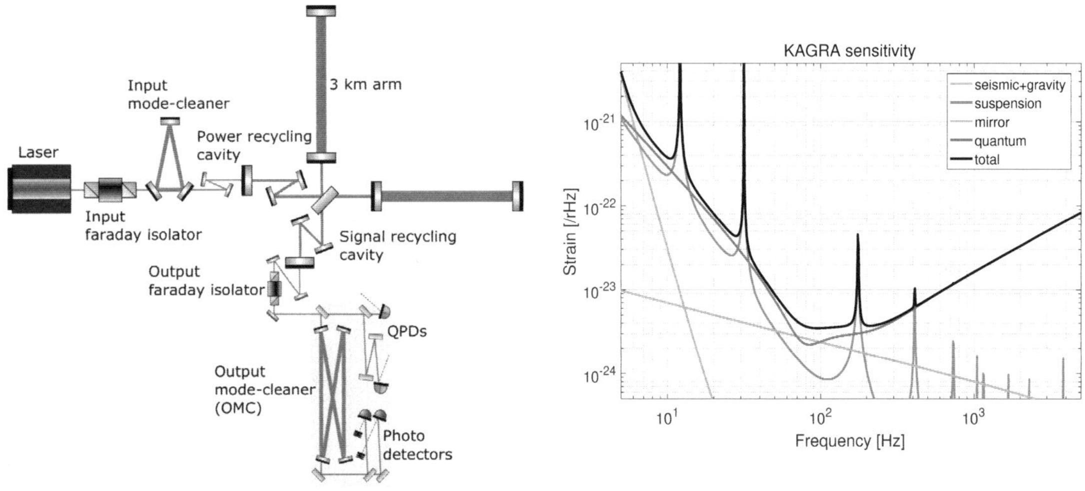

Figure 1. Left: Schematic view of the KAGRA interferometer. The OMC is located at the AS port, in the same vacuum chamber as DC photo detectors. Right: The latest noise budget of KAGRA. The observation range for binary neutron stars is 153 Mpc.

2. Output mode-cleaner

2.1. Optical requirement

The OMC filters out radio-frequency (RF) control sidebands and spatial higher order modes of the carrier light as the carrier light and the gravitational wave signal transmits through it. In order for the sensitivity deterioration in the OMC to be within 5%, a possible breakdown of an acceptable degradation is as follows: the signal loss is to be 2% or less, the residual spatial higher order modes are to be 10 μW or less, and the residual RF sidebands are to be 20 μW or less. The optical parameters of the OMC determined from results of FINESSE [5] simulations are as follows: (i) round-trip cavity length is 1.5 m, (ii) Gouy phase shift is 55.4 deg, (iii) finesse is 780, and (iv) reference field light is 1 mW [6].

2.2. Suspension requirement

Even though all the OMC optics are on a single breadboard, seismic vibration of the OMC system can cause several problems. (i) The signal transmission can be reduced due to optical mode mismatching. (ii) Noise can increase due to modulation to the reference light field. We simulated the degradation of the sensitivity with tilt and mis-centering vibrations of the OMC using FINESSE with artificial mirror maps and calculated the vibration isolation requirement (see Fig. 2, right). We found that the requirement in the horizontal direction is more severe than the vertical direction: the required isolation ratio is 1/100 at 20 Hz. Concerning the vertical-horizontal coupling, we decided to apply the same requirement as the horizontal direction also in the vertical direction. As a result, the OMC should be suspended by three single-stage pendula, each of which supported by a blade spring. (iii) Scattered light may be re-injected from the OMC optics to the main interferometer. Since the solid angle of the KAGRA OMC is 1e-6 sr and the bidirectional reflectance distribution function is 0.1, the backscattered light of the OMC is 0.1 ppm, which is small enough to not disturb the sensitivity of the interferometer. (iv) The signal transmission can be reduced due to an OMC cavity length change. Deformation of the breadboard at its mechanical resonances may change the OMC optical length. We calculated the resonant frequencies using a Finite Element Method (FEM) simulation and determined the silica board size to be 200×500×50 mm, which weighs 11 kg; the lowest mechanical mode is then 1145 Hz.

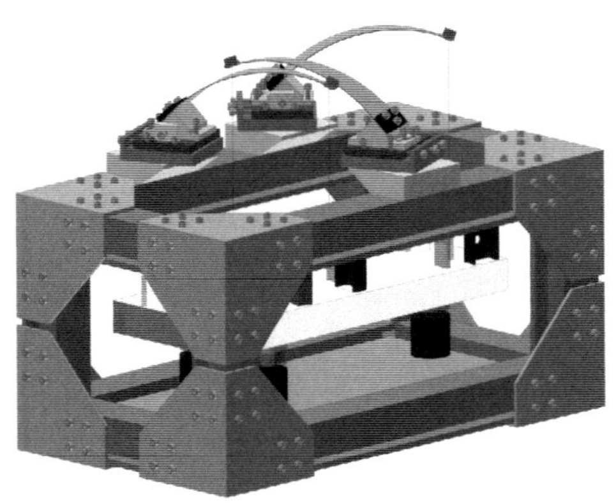

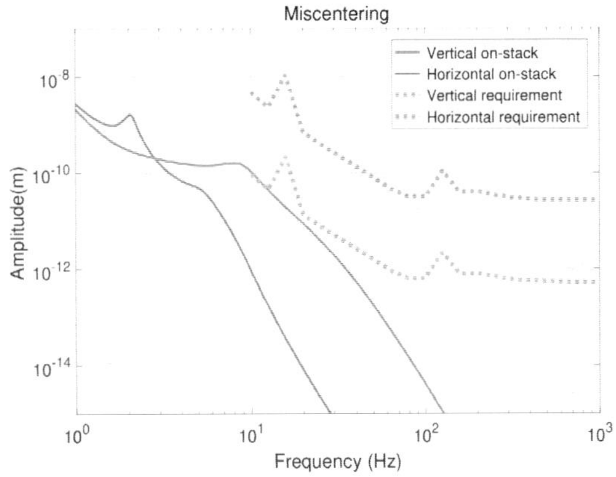

Figure 2. Left: The design drawing of the KAGRA OMC system. The OMC is suspended by three single-stage pendula, each of which supported by a blade spring, and the frame consisting of I-beam is connected with corner joining plates to enhance rigidity. Right: Result of vibration isolation requirement for the KAGRA OMC. The solid lines indicate the noise level due to seismic vibration and the dashed lines indicate the vibration isolation requirements; blue for vertical and orange for horizontal. In KAGRA, the noise level should be reduced to 1/100 or less of the requirement by the anti-vibration device. It is also necessary to consider the vertical-horizontal coupling.

3. Suspension system for the KAGRA OMC

We adapt three blade springs made of MAS-1 (Yield stress: 1.9 GPa) for the OMC vertical seismic isolation. Since the OMC breadboard is not rotationally symmetric, it is essential to design the blade spring to reduce the vertical-horizontal coupling at the working point so that a vertical bounce will not cause a horizontal motion.

The blade spring for the KAGRA OMC is a flat cantilever without load, which is to be curved with a load in a proper radius of curvature, is used to eliminate the vertical-horizontal coupling [7] (see Fig. 3, left). The force applied to the bent blade is expressed by the following formula:

$$\sigma = \frac{Et}{2R}, \tag{1}$$

where σ is the working stress, E is Young's modulus, t is the thickness of the blade, and R is the radius of the bent blade spring. The width of the curved blade spring needs to be precisely described by the following equation:

$$w(x) = \frac{6mgR}{\sigma t^2} \sin\left(\frac{x}{R}\right), \tag{2}$$

where m is the mass of the load and g is the gravitational acceleration. We define the top of the blade spring as $x = 0$ and the bottom as $x = l$ where l is the length of the spring. The resonant frequency of the blade spring is then:

$$f_0 = \frac{1}{2\pi}\sqrt{\frac{g}{R(\theta \sin\theta + \cos\theta - 1)}}, \tag{3}$$

where θ is the installation angle of the blade spring. We selected the parameters so that (i) the safety factor is greater than 3, (ii) the size to apply to the KAGRA OMC system is optimal, and

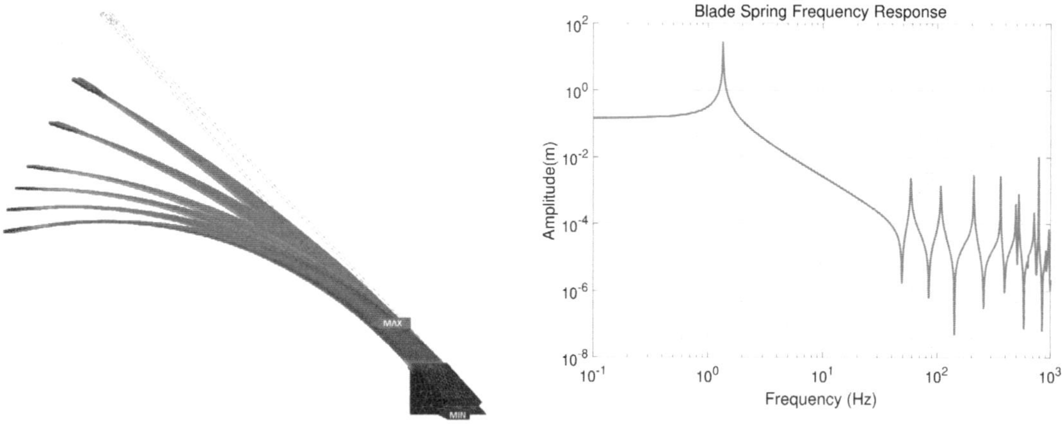

Figure 3. Left: Picture of zero vertical-horizontal coupling blade spring for the KAGRA OMC suspension. Right: Result of a transfer function analysis. We calculated the frequency response of the blade spring using ANSYS, the FEM engineering software.

(iii) it does not cause a horizontal motion to vertical motion at the working point. We reinforced the lateral width to meet the requirement of safety factor using a nonlinear FEM simulation and calculated the frequency response of the blade spring (see Fig. 3, right). Table 1 shows the final design values of the suspension system.

Table 1. Parameter list of the KAGRA OMC blade spring

Total mass	13 kg	Width	52.7mm @bottom
Number of blade	3	Thickness	1.3 mm
Material	MAS-1	Initial angle	45 deg
Yield stress	1.9 GPa	Radius of blade	190 mm
Young's modulus	182 GPa	Working stress	661 MPa
Length	220 mm	Resonant freq.	2.23 Hz

4. Summary and Acknowledgement

The OMC plays an important role in achieving the target sensitivity of KAGRA. In this proceeding, we described the optical and suspension design of the KAGRA OMC according to the requirement calculated with a modal model simulation using artificial mirror maps.

This work was supported by JSPS Specially Promoted Research 26000005, JSPS Core-to-Core Program, the joint research program of the ICRR, and the Mitsubishi Foundation. The authors would like to express our sincere appreciation to our colleagues in the LIGO group.

References

[1] Einstein A 1916 *Annalen der Physik* **49** 769
[2] Abbott B P *et al.* 2016 *Phys. Rev. Lett.* **116** 061102
[3] Somiya K (for the KAGRA Collaboration) 2012 *Class. Quantum Grav.* **29** 124007
[4] Fricke T T *et al.* 2012 *Class. Quantum Grav.* **29** 065005
[5] Freise A, Brown D, and Bond C 2013 arXiv:1306.2973
[6] Kumeta A, Bond C, and Somiya K 2015 *Opt. Rev.* **22** 149
[7] Smith M and Amato-Grill J 2010 *LIGO report* T0900324-v2

A Multi-Step Approach to Assessing LIGO Test Mass Coatings

Lamar Glover[1], Michael Goff[1], Seth Linker[1], Joshua Neilson[1], Jignesh Patel[1], Innocenzo Pinto[2], Maria Principe[2], Ethan Villarama[1], Eddy Arriaga[1], Erik Barragan[1], Shiuh Chao[3], Lara Daneshgaran[1,4], Riccardo DeSalvo[1,2], Eric Do[1], Cameron Fajardo[1]

[1]*California State University, Department of Physics and Astronomy, 5151 State University Drive, Los Angeles CA 90032, USA*
[2]*University of Sannio at Benevento, 82100 Benevento, Italy and INFN (Sezione di Napoli), Italy*
[3]*National Tsing Hua University, Institute of Photonics Technologies, Taiwan, R.O.C.*
[4]*Crescenta Valley High School, 2900 Community Avenue, La Crescenta, CA 91214*
Now at University of California, San Diego, 9500 Gilman Drive, La Jolla, CA 92093

e-mail address: lglover4@calstatela.edu

Abstract. Photographs of the LIGO Gravitational Wave detector mirrors illuminated by the standing beam were analyzed with an astronomical software tool designed to identify stars within images, which extracted hundreds of thousands of point-like scatterers uniformly distributed across the mirror surface, likely distributed through the depth of the coating layers. The sheer number of the observed scatterers implies a fundamental, thermodynamic origin during deposition or processing. If identified as crystallites, these scatterers would be a possible source of the mirror dissipation and thermal noise, which limit the sensitivity of observatories to Gravitational Waves. In order to learn more about the composition and location of the detected scatterers, a feasibility study is underway to develop a method that determines the location of the scatterers by producing a complete mapping of scatterers within test samples, including their depth distribution, optical amplitude distribution, and lateral distribution. Also, research is underway to accurately identify future materials and/or coating methods that possess the largest possible mechanical quality factor (Q). Current efforts propose a new experimental approach that will more precisely measure the Q of coatings by depositing them onto 100 nm Silicon Nitride membranes.

1. Introduction

The mirror test masses of the LIGO Gravitational Wave detector [1][2] have been specifically designed and constructed with multi-layered interference coatings [3] deposited via ion-beam deposition to minimize optical absorption, mirror thermal noise [4][5] and light scattering [6]. As precision increases, various noises become more prominent in the detected signal of LIGO. Identifying and eliminating noise sources become vital. At the level of 10^{-18} m, the surfaces of the test mass mirrors move under the effects of thermal excitation. Mitigating thermal noise at this level is crucial for detecting weaker gravitational wave signals from farther away in the Universe. In addition to mirror thermal noise, the scattering of the LIGO stored beam light from the mirror surface also contributes to the noise budget. Light power decreases and the quantum squeezing of the stored beam is affected by the scattering [7]. Scattering is attributed to irregularities in the refractive index of the coating layers, caused by either voids or denser crystallites within the coating layers. A reduction of

Content from this work may be used under the terms of the Creative Commons Attribution 3.0 licence. Any further distribution of this work must maintain attribution to the author(s) and the title of the work, journal citation and DOI.
Published under licence by IOP Publishing Ltd

IOP Conf. Series: Journal of Physics: Conf. Series **957** (2018) 012010 doi:10.1088/1742-6596/957/1/012010

the current levels of scattered light could also potentially increase the current sensitivity of LIGO, by allowing stronger squeezed light in Advanced LIGO [7]. The research efforts at Cal State LA are focused on detecting and identifying scatterers within the LIGO mirror coatings for these purposes.

Figure 1: November 2015 Photograph of an End Test Mass at LIGO Hanford with an exposure time of 1.3 seconds. (Note: The rings and faint straight lines that appear in the background on the right side of the image are components of the mirror support structure).

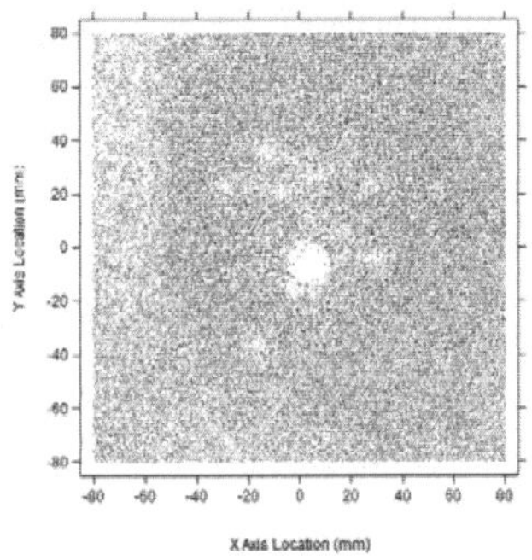

Figure 2: DAOPHOT-detected scatterers with 0.763 second exposure. Smear due to residual surface impurities limits DAOPHOT's detection efficiency in some areas that appear blank.

Until recently, a portion of the observed scattered light was attributed to things such as residual surface impurities deposited on the mirrors during implementation accesses. It was believed that many could be removed with swiping, First Contact and other clean room practices, even if completely clean mirrors could and cannot be achieved. However, using the astronomical image software suite DAOPHOT [8][9] to analyze in-situ photographs of LIGO end test masses, hundreds of thousands of weaker, point-like scatterers have been detected, uniformly distributed across the mirror surface, ignoring the smear of residual surface impurities on the mirror. The amplitude distribution of the observed scatterers implies that they extend through the depth of the coating while their sheer number implies a fundamental, thermodynamic origin pointing towards the presence of crystallites.

Motivated by these observations, the Cal State LA LIGO Research Group developed a scatterer mapping technique [10] capable of sufficient depth resolution not only to identify the layer (high or low-refraction index) hosting the scatterer, but also the distribution within each layer. A density growing towards the top of the layer would signal the presence of crystallites, which are observed to evolve into crystalline columnar growth when layers are thicker. [11] Voids are expected to be more prominent at the coating layer boundaries, near the interface to the previous glass, which has a different atomic spacing and structure (SiO_2 vs. Ti_2O_5 for example). [12] This experiment could potentially distinguish between the two possibilities.

A second effort, in collaboration with Tsing Hua, is working towards the measurement of the quality factor of new coating samples. Further research would follow to determine if there is any correlation between the Q-factor measurements and detected scatterer count and size within the coating samples.

2. Feasibility of mapping scatterers within LIGO coating layers

The detection of hundreds of thousands of sub-wavelength, weak scatterers in the LIGO mirror coatings has served as an impetus for further research. While observing the scatterers with the 100 kW light power of LIGO is informative, it cannot be used to improve the current sensitivity of the device since the mirrors are already installed and cannot be modified or replaced during this stage of LIGO operation. Also, most light is reflected in the first few layers and the scattered light is further attenuated by reflection in the way that the observed scattering effectively explores only the first few

IOP Conf. Series: Journal of Physics: Conf. Series **957** (2018) 012010 doi:10.1088/1742-6596/957/1/012010

layers. To truly learn more about the proliferation of the scatterers throughout the coating layers, in-situ photographs would not be sufficient.

As a follow-up to the previously explained analysis, it is necessary to produce a complete mapping of scatterers in a sample including the amplitude distribution, the lateral distribution (to check for uniformity), and the depth distribution. For best guidance of the coating process, it would be ideal to achieve depth sensitivity sufficient not only to know if the scatterers belong to a high refraction index or low refraction index quarter wavelength layer, but also to determine if they cluster near the bottom or top of the layer, or if they are uniformly distributed.

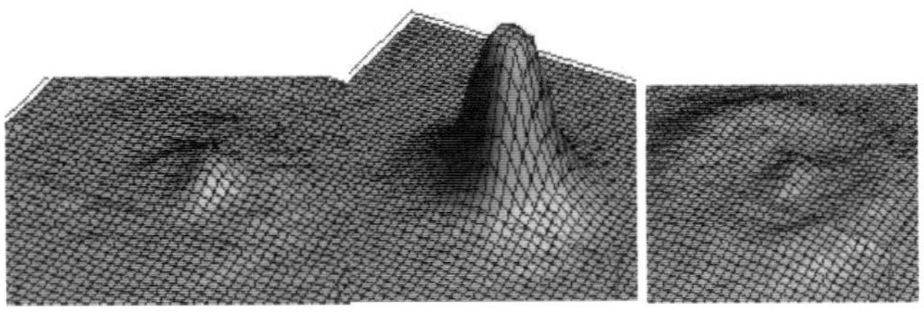

Figure 3: Dark field, 20x 3D images of ROI containing a scatterer moving from out of focus, to in focus, to out of focus as the scan progresses. Note the Airy function rings present when out of focus. In the center the image is fully focused and the Airy function rings have collapsed into the central peak, which does not significantly change width. The pictures were taken at roughly 15 steps apart.

As an initial feasibility study, a microscope with both bright and dark field was used for observations on a 25 mm diameter commercial dielectric mirror and images of diffraction-limited scatterers were found in the mirror. When properly focused, scatterers appear as bright dots on dark field or dark dots with bright field in microscope observation.

Out-of-focus scatterers are not visible or appear as blurred dots. As the focal plane is scanned through the coating layers, scatterers first appear as fuzzy blotches, then they show as an Airy function with rings. When in complete focus, the rings coalesce into the central peak that becomes sharp and bright. As the scan progresses, the sequence inverts, until each and all scatterers disappear. Three points of a sequence is illustrated in figure 3.

We take advantage of this sequence to measure the depth of each scatterer in CCD microscope images, by fitting the shape and amplitude of the Airy function images as the position of the focal plane goes through a micrometrical scan through the coating's thickness.

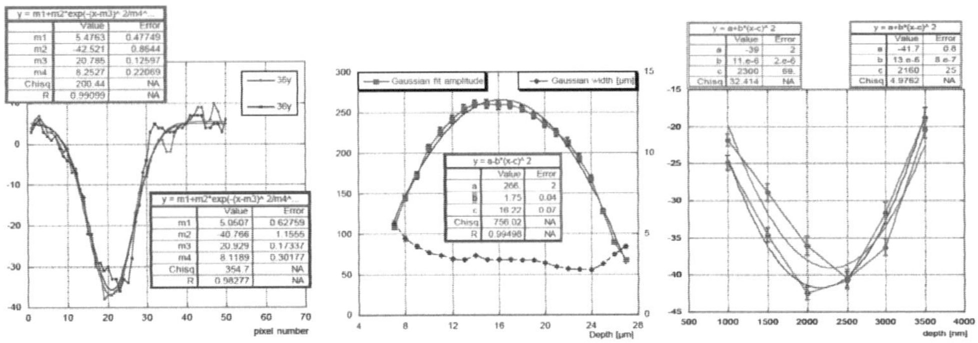

Figure 4: Left: Gaussian fit to two dark field Airy function slices, whose amplitude (m1) is used to find the scatterer depth. Center: amplitude vs. depth fitting with dark field (20x magnification) used to determine the depth (C). Right: Two bright field (50x magnification) amplitude vs. depth fittings, a slightly better depth resolution is achieved.

Dielectric mirrors reflect only under specific conditions, i.e. at very specific wavelengths and angles so that the light returning from each layer is in phase. With the cone of illumination light in the microscope, with both bright or dark field, light comes in at all angles. In addition, the white light used contains all wavelengths of visible light. The multiple angles and wavelengths ensure that, unlike the case of the stored beam illuminated LIGO mirrors, the light can reach all layers and scatterers can be detected at all depths.

Before proceeding, it is useful to note that, due to the refraction index within the coating, the focusing beams are compressed in depth by Snell's law. Each movement of the microscope in air corresponds to lower distance within the sample. The best depth resolution obtained, 30 nm, corresponding to 3% of the 1064 nm LIGO light wavelength, should be divided by the local refraction index, thus yielding even higher accuracy inside the glass if measured in nm, but always 3% of the wavelength. This resolution allows for identification of not only the layer in which the scatterer is located, but even whether it is located near the top, middle, or bottom of each layer: crucial information in understanding the physics at play.

The worse resolution obtained is ~100nm, while the best has been ~30 nm. When considering calculations that account for the refraction indexes of the coating layers, the achieved resolution of ~ 50 nm allows clear identification of the host layer.

As the current system evolves, there should be complete automation in the Z-axis as well as a X-Y axes raster scan, allowing for better resolution and comprehensive analysis of coating samples in order to detect scatterers with greater detail.

3. Feasibility of measuring coating Q-factors with low dilution factors using SiN membranes

While this microscopic search technique could prove to be useful in detecting and locating scatterers in greater detail, in order to determine if there is a correlation between scatterers and thermal noise, it is necessary to also measure the quality factor of different coatings. This, followed by microscopic analysis of these same coatings, would provide information on the quantity and distribution of scatterers within these samples and determine if there is a correlation between scatterers and coating q-factors. Currently, there are efforts being made by the Cal State LA LIGO Research Group, in collaboration with Tsing Hua, moving towards just that.

A common method of q-factor assessment involves coating a ~50 μm thick cantilever beam and clamping it at one end. The substrate, in this method, is ~100 times thicker than the coating which yields D on the order of 10^{4}. The cantilever system has an additional problem due to the clamping mechanism. It has been observed [13] that losses from clamping are significant and can lead to unpredictable measurements. Just unclamping and re-clamping the cantilever introduces perturbations sufficient to severely affect the measurements of Q.

A different method has been devised to avoid clamping the system altogether. This is the GeNS (Gentle Nodal Suspension) and involves balancing a thin (~1 mm) fused silica disk on a small diameter sapphire bead. The center of the disk is a nodal point of the even modes that, therefore, can oscillate with virtually no losses [14]. There are very limited holding losses but in this case the substrate is now ~10,000 times thicker than the coating [15] and a value for D on the order of 10^{-8} contributes to an uncertainty of the coating's Q.

Silicon nitride membranes are a new material used to fabricate windows for electron microscopes. In some ways, it is ideal as a substrate for the coatings. They are grown over a silicon wafer substrate, where a window then is etched at the center to free up the membrane with a strong silicon frame. They have been shown to possess quality factors on the order of 10^{7}, and are strong enough to hold atmospheric pressure even at thicknesses of 100 nm. Due to their minimal thickness and to their strength, a ~1:1 coating to substrate thickness ratio can be achieved.

The basic experimental idea is to coat the front and back of the membranes with the material to be tested, hold the frame with as little clamping losses as possible, electro-statically excite the individual modes and monitor the oscillation ring-down with an optical lever system to make precise measurements of the Q of the coatings.

The main drawback of using SiN windows as substrates is that the loss angle of the coating materials should be measured within the sensitivity range of the gravitational wave detectors, ~1 to 1000 Hz, because the material Q factor is not necessarily constant across all frequencies. The resonant frequency of typical SiN windows is between ten and hundreds of kHz, well outside the frequency region of interest. Depositing the coatings adds mass linearly with thickness, which reduces the resonant frequency but also increases the moment of inertia with the cube of the thickness. Thus, neglecting the tensioning contribution, the resonant frequency can be expected to grow (more or less) linearly with thickness, moving even further from the region of interest.

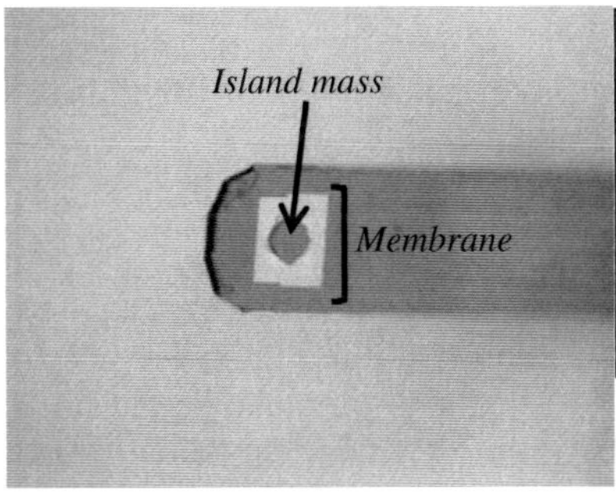

Figure 5: Pictured is an actual sample in our possession that was manufactured by Chao Shiuh's group at National Tsing Hua University. The membrane and island mass are at the end of a long stem to prevent damage during shipping. The island mass has a diameter ~380 microns. The exposed membrane area is ~890 microns, with a thickness of ~100nm. We will glue the entire stem to our fused silica stage using the UV activated epoxy, with minimal losses.

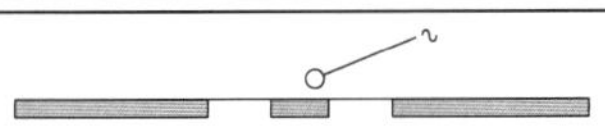

Figure 6: An electrode excites the coated membrane at the desired frequency. We place it off-center from the axis to provide the torque required for higher order modes.

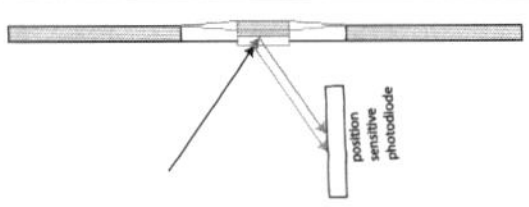

Figure 7: First mode of resonance, and we see how the motion is measured by a laser reflected onto a position sensing photodiode.

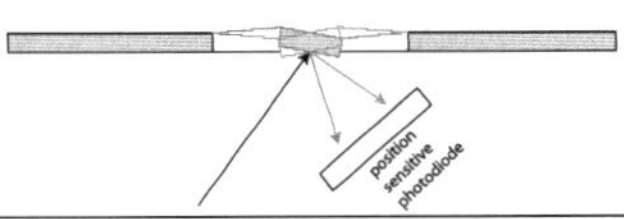

Figure 8: The second resonant (tilt) mode

To reduce the resonant frequency of the membrane, we can have it manufactured where an island of silicon is left in the center. The island's mass will affect the frequency with the usual formula

$$\omega = \sqrt{\frac{k}{m}} \qquad \text{Eq 1}$$

while the width of the membrane around the island determines the effective restoring force constant k.

Considering that the silicon is typically 10,000 times thicker than the SiN membranes, resonant frequency reductions on the order of 100 may be achievable, thus bringing the measurement much closer to the frequency region of interest.

To mitigate the clamping problem the frame will be bonded to a fused silica substrate, which in turn will be clamped in the UHV apparatus, possibly suspended from a vibration isolation stage.

4. Conclusion

The ongoing aspects of research that are underway all complement one another. With the detection of scatterers from in-situ images of LIGO, the Cal State LA LIGO Research Group was motivated to take a more in-depth look at detecting scatterers within samples by means of microscopic imaging. Also, determining if there is a correlation between the number of scatterers within a sample and its determined Q factor has the potential to make a big impact on how we understand the coating materials used on the test mass mirrors. There is the possibility that all these different avenues of complementary research could paint an intricate mosaic, giving us a better understanding of the coating materials that are being utilized. This could potentially help increase the sensitivity and detection rate of GW events, making better astrophysics.

References

[1] Aasi, Junaid, et al. "Advanced LIGO." *Classical and Quantum Gravity* 32.7 (2015): 074001.
[2] Abbott, B. P., et al. "GW151226: Observation of gravitational waves from a 22-solar-mass binary black hole coalescence." *Physical Review Letters* 116.24 (2016): 241103.
[3] Harry, Gregory, Timothy P. Bodiya, and Riccardo DeSalvo, eds. *Optical coatings and thermal noise in precision measurement.* Cambridge University Press, 2012.
[4] Villar, Akira E., et al. "Measurement of thermal noise in multilayer coatings with optimized layer thickness." *Physical Review D* 81.12 (2010): 122001.
[5] M. Principe, "Reflective Coating Optimization for Interferometric Detectors of Gravitational Waves," *Optics Express.* 23.9 (2015) 10938-10956.
[6] Harry, Gregory M., and LIGO Scientific Collaboration. "Advanced LIGO: the next generation of gravitational wave detectors." *Classical and Quantum Gravity* 27.8 (2010): 084006.
[7] John Miller et al., "Prospects for doubling the range of Advanced LIGO," *Physical Review D* 91 (2015): 062005-2.
[8] Glover, L., et al., "Observation of a Large Population of Optical Scatterers in the Advanced LIGO Mirrors," LIGO Scientific Collaborative Database LIGO-P1600325-v4 (2017).
[9] Glover, L., et al. "Optical scattering measurements and implications on thermal noise in Gravitational Wave detectors test-mass coatings," *Physics Letters A* (2017), http://dx.doi.org/10.1016/j.physleta.2017.05.050
[10] Daneshgaran, Lara, Michael Milovich-Goff, Lamar Glover, Riccardo DeSalvo, "Feasibility study of mapping scatterer within the depth of Dielectric coatings," LIGO Scientific Collaborative Database LIGO-T1700458-v1 (2017).
[11] Huang-Wei Pan, Shun-Jin Wang, Ling-Chi Kuo, Shiuh Chao, Maria Principe, Innocenzo M. Pinto, and Riccardo DeSalvo, "Thickness-dependent crystallization on thermal anneal for titania/silica nm-layer composites deposited by ion beam sputter method." Optics Express, 22.24 (2014): 29847-29854
[12] Steven D. Penn, Gregory M. Harry, Andri M. Gretarsson, Scott E. Kittelberger, Peter R. Saulson, John J. Schiller, Joshua R. Smith, Sol O. Swords. "High quality factor measured in fused silica ", Review of Scientific Instruments 72 (2001): 3670-3673
[13] Numata, Kenji, "Intrinsic losses in various kinds of fused silica," Class. Quantum. Grav. 19:7 (2002)
[14] Granata, M., et al, "Correlated evolution of structure and mechanical loss of a sputtered silica film," (2017): arXiv:1706.02928
[15] Zwickl, B. M., et al., "High quality mechanical and optical properties of commercial silicon nitride membranes," Appl. Phys. Lett. 92 (2008): 103125

A new class of compact high sensitive tiltmeter based on the UNISA folded pendulum mechanical architecture

Fabrizio Barone[1], **Gerardo Giordano**[2]

[1] Department of Medicine, Surgery and Dentistry *Scuola Medica Salernitana*, University of Salerno, Via Salvator Allende, Baronissi (Salerno) 84081, Italy

[2] University od Salerno, Via Giovanni Paolo II, Fisciano (Salerno) 84084, Italy

E-mail: `fabrizio.barone@unisa.it`

Abstract. We present the Extended Folded Pendulum Model (EFPM), a model developed for a quantitative description of the dynamical behavior of a folded pendulum generically oriented in space. This model, based on the Tait-Bryan angular reference system, highlights the relationship between the folded pendulum orientation in the gravitational field and its natural resonance frequency. Tis model validated by tests performed with a monolithic UNISA Folded Pendulum, highlights a new technique of implementation of folded pendulum based tiltmeters.

1. Introduction

The innovative architecture of the UNISA class of folded pendulums (based on the classical Watt's Linkage) allow the implementation of compact, robust and light sensors, characterized by high sensitivity and large band, especially in the low frequency region, configurable as seismometers, velocimeters and/or accelerometers, capable to operate also in harsh, ultra-high-vacuum and cryogenic environments (UHV) [1, 2, 3, 4]. Implemented in different materials and with different readouts to satisfy the requirements of a large variety of applications aimed both to control and monitoring (earthquake engineering, geophysics, civil engineering, etc.) [5, 6, 7, 8, 9], these devices appear to be very interesting also as sensors for applications to interferometric detectors of gravitational waves of the present and next generation [10]. But despite the high quality of the present implementations, a relevant problem still remains unsolved: the output signal is always a combination of linear and angular signals [1]. This ambiguity requires relevant assumptions on the output signal, although often very close to the real situation, like, for example, that one of the two signals is much higher than the other one or that the two signals have different bands [11, 12].

In this paper, we demonstrate that it is possible to obtain the angular signal by reinterpreting the output signal on the basis of the Extended Folded Pendulum Model (EFPM). This model, developed for a quantitative description of the dynamical behavior of a folded pendulum generically oriented in the gravitational field, highlights the relationship between its orientation and natural resonance frequency. The *EPFM*, based on the Tait-Bryan angular reference system and validated by the results of tests performed with a monolithic UNISA Folded Pendulum, opens the way to new interesting applications of the folded pendulum as tiltmeter.

Content from this work may be used under the terms of the Creative Commons Attribution 3.0 licence. Any further distribution of this work must maintain attribution to the author(s) and the title of the work, journal citation and DOI.
Published under licence by IOP Publishing Ltd

2. Generalized Folded Pendulum Model

Although the folded pendulum dynamical behavior can be described with enough accuracy only with numerical approaches, the Lagrangian two-dimensional analytical model of Bertolini [14], based on the simplified Liu et al. model [13] and generalized by Barone et al. [1, 2] to describe the dynamics of a generically oriented folded pendulums on a vertical plane, is generally sufficient for a synthetic analysis. The generalized model requires the definition of two reference systems, the first, xyz, integral to the folded pendulum, characterized by the x axis parallel to the direction of motion of the central mass and by the y axis parallel to the hinges (or to the flexures) rotation axes, while the second one, XYZ, is characterized by the plane XY perpendicular to the direction of the gravitational acceleration vector, $\vec{g}$ (Figure 1). In the generalized folded pendulum model [1, 2], the xyz reference system can rotate around the Y ($Y \equiv y$) axis of a generic angle β, positive for a counterclockwise rotation, so that an horizontal folded pendulum is described by a rotation angle $\beta = 0\,rad$, while a vertical one by $\beta = \pi/2\,rad$. This model, schematized in Figure 1 for $\beta = 0$, consists of two vertical arms of equal length, l_p, connected to one side to a single support (frame) by means of two hinges, forming a simple pendulum of mass m_{p_s} and an inverted pendulum of mass m_{p_i}. The two pendulums masses are concentrated in their centers of mass, P_s and P_i, respectively, positioned in $l_b = l_p/2$. The other sides of the arms are connected (in C_s and C_i, respectively) to a bar of mass, m_c, and length, l_d, by means of two other hinges. The mass of the central bar is modeled with two equivalent masses m_{c_s} and m_{c_i}, being

$$m_c = m_{c_s} + m_{c_i} \tag{1}$$

whose values depend on the position of the center of mass on the central bar, l_m (being $l_m < l_d$), measured with respect to the pivot point, C_s, according to the relations ($\beta = 0$) [2]

$$m_{c_s} = m_c \left(1 - \frac{l_m}{l_d}\right) \qquad\qquad m_{c_i} = m_c \left(\frac{l_m}{l_d}\right) \tag{2}$$

Taking into account that the UNISA Folded Pendulum is monolithic and that its pivot points are elastic flexure joints, then for a generic orientation ($\beta \neq 0$) the angle describing the central rod rest position, θ_o, is determined by the equilibrium position determined by the forcement due to the component of the equivalent gravitational acceleration along the x-axis direction, g_x, and by the reaction due to the elastic force generated by the folded pendulum flexure joints, and described by the following equation

$$M_{eq} l_c g_x(\beta) - k_\theta \theta_o = 0 \tag{3}$$

being $\vec{g}(\beta)$, the gravitational acceleration vector, expressed by

$$\vec{g}(\beta) = g_x(\beta)\,\vec{i} - g_z(\beta)\,\vec{k} = g\sin\beta\,\vec{i} - g\cos\beta\,\vec{k} \tag{4}$$

A change of orientation, quantified by the rotation angle, β, produces a displacement of the centre of the mass m_c of a quantity Δl_{θ_o}, that, can be expressed at the first order by the trigonometric relation

$$\Delta l_{\theta_o}(\beta) \approx l_c \theta_o(\beta) = \frac{M_{eq} g_{eq_x}(\beta)}{K_{e_{eq}}} \tag{5}$$

that, in turn, produces changes in m_{c_s} and m_{c_i}, now functions of the angle β

$$m_{c_s}(\beta) = m_{c_s} - \frac{m_c}{l_d}\Delta l_{\theta_o}(\beta) \qquad\qquad m_{c_i}(\beta) = m_{c_i} + \frac{m_c}{l_d}\Delta l_{\theta_o}(\beta). \tag{6}$$

The solution of the lagrangian equations including these effects, albeit simplified, shows that

IOP Conf. Series: Journal of Physics: Conf. Series **957** (2018) 012011 doi:10.1088/1742-6596/957/1/012011

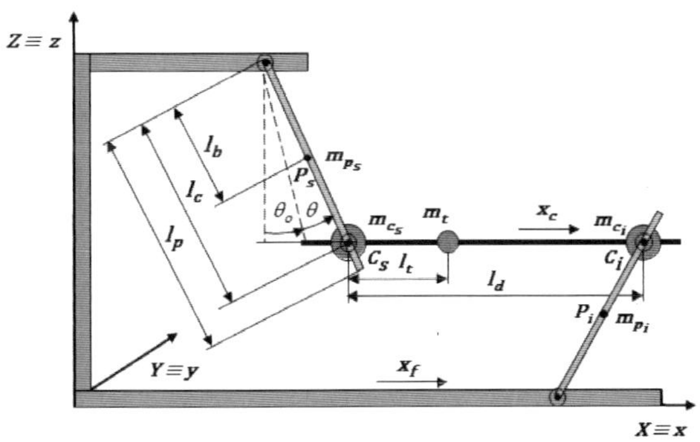

Figure 1. Generalized Folded Pendulum scheme.

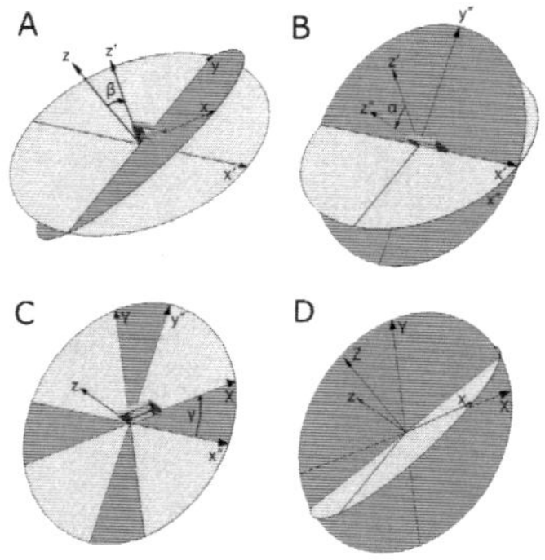

Figure 2. Tait Bryan Rotation Sequence ($\beta\alpha\gamma$).

the folded pendulum can be still effectively described as a second order system [1], with small deflection angles, θ, evaluated around θ_o. In fact, assuming the folded pendulum frame rigidly connected to the ground, that is $x_f(t) = x_g(t)$, being $x_g(t)$ the component of the ground motion parallel to the direction of the inertial mass, then the folded pendulum transfer function is expressed as

$$H(s) = \frac{X_c(s) - X_g(s)}{X_g(s)} = \frac{X_{output}(s)}{X_g(s)} = \frac{-(1 - A_c)\,s^2}{s^2 + \frac{\omega_o}{Q(\omega_o)}s + \omega_o^2} \tag{7}$$

where $Q(\omega_o)$ is the global mechanical quality factor, A_c is the parameter related to the center of percussion effects and ω_o is the angular frequency [1], expressed, in absence of losses, by

$$\omega_o(\beta) = \sqrt{\frac{\left[(m_{p_s} - m_{p_i})\frac{l_p}{2l_c} + (m_{c_s} - m_{c_i}) - 2m_c\frac{M_{eq}g_x(\beta)}{K_{e_{eq}}l_d}\right]\frac{g_z(\beta)}{l_c} + \frac{k_\theta}{l_c^2}}{(m_{p_s} + m_{p_i})\frac{l_p^2}{3l_c^2} + (m_{c_s} + m_{c_i})}} = \sqrt{\frac{K_{g_{eq}}(\beta) + K_{e_{eq}}}{M_{eq}}} \tag{8}$$

being k_θ the sum of the contributions of the elastic constants of all the folded pendulum flexures [1]. $K_{e_{eq}}(\beta)$ and $K_{g_{eq}}(\beta)$ are instead the equivalent elastic constants of the folded pendulum, modeled as a spring-mass equivalent oscillator, defined, respectively, as global equivalent elastic constant and gravitational equivalent elastic constant,

$$K_{e_{eq}(\beta)} = \frac{k_\theta}{l_c^2} \qquad K_{g_{eq}}(\beta) = \left[(m_{p_s} - m_{p_i})\frac{l_p}{2l_c} + (m_{c_s} - m_{c_i}) - 2\frac{m_c}{l_d}\frac{M_{eq}g_x(\beta)}{K_{e_{eq}}}\right]\frac{g_z(\beta)}{l_c} \tag{9}$$

while M_{eq} is the equivalent mass, defined as

$$M_{eq} = (m_{p_s} + m_{p_i})\frac{l_p^2}{3l_c^2} + (m_{c_s} + m_{c_i}) = (m_{p_s} + m_{p_i})\frac{l_p^2}{3l_c^2} + m_c \tag{10}$$

3. The Extended Folded Pendulum Model

In the $EFPM$ the folded pendulum orientation with respect to the local horizontal is quantified by the relative orientation between the xyz and XYZ reference systems, defined by the Tait-Bryan angles: a rotation angle around x axis (roll angle, α), a rotation angle around y axis (pitch angle, β) and a rotation angle around z axis (yaw angle, γ). The rotation sequence, chosen for the *EFPM* model [15] is the $\beta\alpha\gamma$ sequence (pitch-roll-yaw), shown in Fig. 2 and described by the following rotation matrix:

$$R = \begin{pmatrix} \cos\gamma\cos\beta + \sin\gamma\sin\alpha\sin\beta & \sin\gamma\cos\alpha & -\cos\gamma\sin\beta + \sin\gamma\sin\alpha\cos\beta \\ -\sin\gamma\cos\beta + \cos\gamma\sin\alpha\sin\beta & \cos\gamma\cos\alpha & \sin\gamma\sin\beta + \cos\gamma\sin\alpha\cos\beta \\ \cos\alpha\sin\beta & -\sin\alpha & \cos\alpha\cos\beta \end{pmatrix} \tag{11}$$

being the folded pendulum horizontally positioned when $\alpha = 0$, $\beta = 0$. In the same, way, the gravitational acceleration vector, $\vec{g}$, is expressed in the reference system integral to the folded pendulum, xyz, as

$$\vec{g} = \begin{pmatrix} \cos\gamma\,\sin\beta - \sin\gamma\,\sin\alpha\,\cos\beta \\ -\sin\gamma\,\sin\beta - \cos\gamma\,\sin\alpha\,\cos\beta \\ -\cos\alpha\,\cos\beta \end{pmatrix} g \tag{12}$$

being $(0, 0, -g)$ the acceleration gravity vector in the XYZ reference frame. Therefore, the gravitational acceleration can be split into three components changing according to the folded pendulum orientation with respect to the XYZ reference system. In particular, for a folded pendulum oriented as shown in Fig. 1, constrained to maintain $\gamma = 0$ the components of the gravitational acceleration along the x-axis, y-axis and z-axis, are respectively

$$g_x(\beta) = g\sin\beta \qquad g_y(\alpha,\beta) = -g\sin\alpha\cos\beta \qquad g_z(\alpha,\beta) = -g\cos\alpha\cos\beta \tag{13}$$

where $g_x(\beta)$ acts along the central mass direction of motion, generating a force opposed by the elastic reaction of the joints; $g_y(\alpha,\beta)$ acts along the direction perpendicular to the motion of the central mass, forcing the joints along a direction parallel to their rotation axes and generating their reaction as constraints; $g_z(\alpha,\beta)$ is the effective component of the gravitational acceleration to used in the simplified Lagrangian model.

The change of orientation of the folded pendulum has a direct effect on its potential energy. The solution of the new lagrangian equations lead to a more general expression or the angular frequency, ω, that now depends on both α and β angles, that is

$$\omega_o(\alpha,\beta) = \sqrt{\frac{\left[(m_{p_s} - m_{p_i})\frac{l_p}{2l_c} + (m_{c_s} - m_{c_i}) - 2\frac{m_c}{l_d}\frac{M_{eq}g}{K_{eeq}}\sin\beta\right]\frac{g}{l_c}\cos\alpha\cos\beta + \frac{k_\theta}{l_c^2}}{(m_{p_s} + m_{p_i})\frac{l_p^2}{3l_c^2} + (m_{c_s} + m_{c_i})}} \tag{14}$$

being $K_{g_{eq}}$ now function of both the roll angle, α, and of the pitch angle, β.

The most evident consequence is that a folded pendulum angular displacement determines a change of its natural resonance frequency, described, for $\gamma = 0$, by Eq. 14 for small, static or slowly variable angular displacements.

In figure 3 the resonance frequency for a standard folded pendulum is shown as a function of both roll angle, α and pitch angle, β. Let us now analyze two limit cases: angular rotation in presence of only roll ($\beta = 0$) (i.e. rotation around the x-axis) and angular rotation in presence of the only pitch ($\alpha = 0$ (i.e. rotation around the y-axis).

In the first case the component of the gravitational acceleration, $g_y(\alpha, 0)$, is always perpendicular to the motion of the central mass and, by definition, totally compensated by the constraint reaction of the joints. Therefore, the resonance frequency is an *even* function the resonance frequency (referred to $\alpha = 0$): a measurement of the resonance frequency allows to obtain the roll angle module, but not the rotation verse. This problem can be solved in different ways [15], the easiest one requiring the use of only one of the two branches of this function, i.e. either $\alpha > 0$ or $\alpha < 0$, positioning the folded pendulum rotated of a known roll angle, $\alpha_o \neq 0$.

In second case, the rotation occurs along the y axis, perpendicular to the motion of the central mass, so that the component of the gravitational acceleration, $g_z(0,\beta)$, perpendicular to the motion of the central mass, changes with the pitch angle, β, generating a *monotone* resonance frequency function with respect to the folded pendulum horizontal orientation.

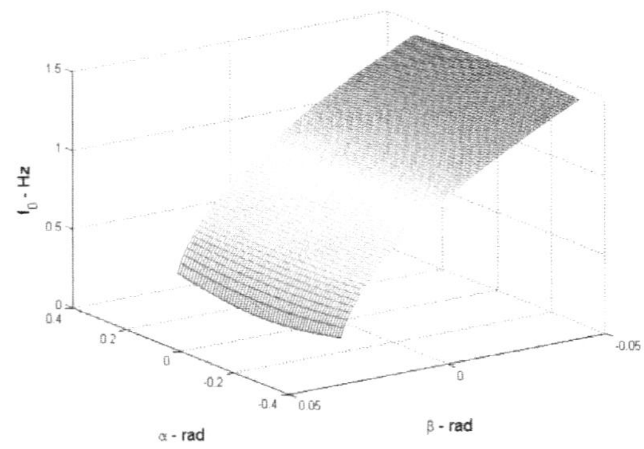

Figure 3. Example of resonance frequency for a typical folded pendulum as function of roll angle, α, and pitch angle, β.

Figure 4. Example of monolithic folded pendulum used for the EFPM test - Model GC12.

4. Tests of the Extended Folded Pendulum Model

The validation of the *EFPM* has been performed using a monolithic UNISA Folded Pendulum of $134\,mm \times 140\,mm \times 40\,mm$ (Model GC12)(Fig. 4), implemented from an Aluminum metal block (Alloy 7075-T6) with precision mechanical milling machining while the eight elliptical joints, characterized by ellipticity ratio of 16/5 and minimum thickness of $100\,\mu m$, have been implemented with electro-discharge machining (EDM). The arms of the pendulum (of length $81.5\,mm$, distant from each other by $102\,mm$ and spaced apart by $6.5\,mm$ from the central mass and the frame) have been designed to minimize mass and inertia momentum without loss of stiffness and symmetry, characterised by a measured quality factor of $Q > 1500$ in air at $\approx f_o = 700\,mHz$ and by an optical lever readout with a position sensing photodiode (PSD) [1]. The signal, acquired with a digital acquisition system at a sampling frequency of $1\,kHz$, pre-filtered by a numerical pass band filter, is then processed to get the resonance sampling frequency. The measurements have been performed in air positioning the prototype on a platform, whose orientation could be changed acting on suitably positioned piezoelectric actuators, optimized to generate independently roll and pitch angles. A step signal has been used to change the angular position of the platform in order to provide, at the same time, also

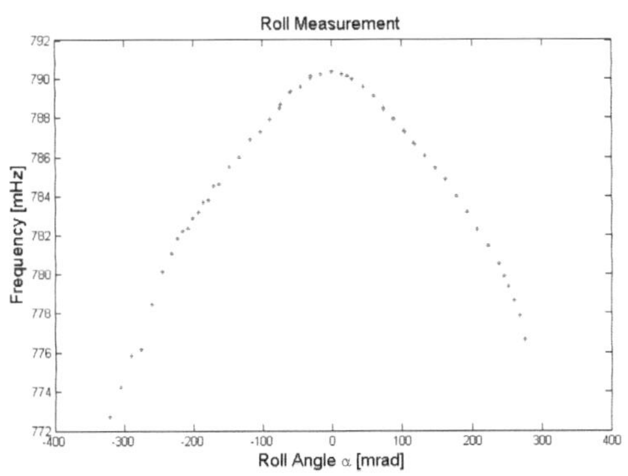

Figure 5. Resonance Frequency vs. roll angle, α, for model GC12.

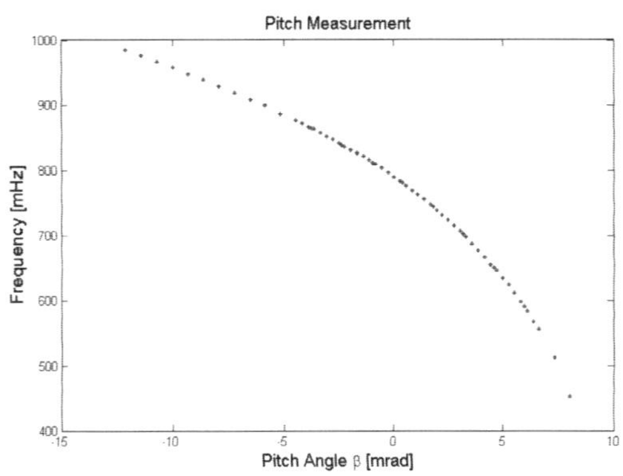

Figure 6. Resonance Frequency vs. pitch angle, β, for model GC12.

the mechanical signal necessary to force the central mass oscillation. Two series of angular displacement measurements are reported, carried out with the folded pendulum positioned as described in the previous sections ($\gamma = 0$), in the above described limit cases of pure roll ($\beta = 0$) and pure pitch ($\alpha = 0$), as reported in Fig. 5 and in Fig. 6. In Fig. 5 the trend of the $GC12$ natural resonance frequency is shown for both positive and negative roll angles, demonstrating that, as predicted by the *EFPM*, the resonance frequency is an *even* function and in very good agreement with the theoretical model. In Fig. 6, instead, the trend of the $GC12$ natural resonance frequency is shown for both positive and negative pitch angles, demonstrating again that, as predicted by the *EFPM*, the folded pendulum natural resonance frequency is a *monotone* function of the pitch angle and in very good agreement with the theoretical model.

5. Conclusions

We have presented a simplified version of the *EFPM*, a model capable to predict the folded pendulum dynamical behavior in the gravitational field, highlighting the relationship between its orientation and the natural resonance frequency. The *EPFM*, based on the Tait-Bryan angular reference system and validated by the results of tests performed with a monolithic UNISA Folded Pendulum, opens the way to new interesting applications of the folded pendulum as tiltmeter.

References

[1] Barone F, Giordano G 2015 *Mechanical Accelerometers*, J. Webster (ed.) Wiley Encyclopedia of Electrical and Electronics Engineering John Wiley & Sons, Inc. doi: 10.1002/047134608X.W8280.

[2] Barone F, Giordano G 2017 *The UNISA folded pendulum: A very versatile class of low frequency high sensitive sensors Measurement* doi:10.1016/j.measurement.2017.09.001.

[3] Barone F, Giordano G *Low frequency folded pendulum with high mechanical quality factor, and seismic sensor utilizing such a folded pendulum* Pat. Num.: IT1394612, EP2452169, JP 5409912, RU 2518587, AU2010269796, US8,950,263, Canada pending.

[4] Barone F, Giordano G, Acernese F *Low frequency folded pendulum with high mechanical quality factor in vertical configuration, and vertical seismic sensor utilizing such a folded pendulum* Pat. Num.: IT1405600, EP2643711, AU201247104, JP5981530, RU2589944, US9256000, Canada pending.

[5] Acernese F, De Rosa R, Giordano G, Romano R and Barone F 2008 *Mechanical monolithic horizontal sensor for low frequency seismic noise measurement. Rev. Sci. Instrum.* **79** 074501 doi:10.1063/1.2943415.

[6] Acernese F, Giordano G, Romano R, De Rosa R and Barone F 2010 *Tunable mechanical monolithic sensor with interferometric readout for low frequency seismic noise measurement Nucl Instrum. and Meth. A* **617** 457 ISSN:0168-9002 doi:10.1016/j.nima.2009.10.112.

[7] Barone F, Giordano G, Acernese F and Romano R 2015 *Watt's linkage based large band low frequency sensors for scientific applications, Nucl Instrum. and Meth. A* **827** 187 doi:10.1016/j.nima.2015.11.015.

[8] Barone F, Giordano G, Acernese F, Romano R 2016 *Tunable mechanical monolithic sensors for large band low frequency monitoring and characterization of sites and structures Proc. SPIE* **9986** 99860C ISBN:9781510603769 doi:10.1117/12.2242080.

[9] Barone F, Giordano G 2017 *Watt's linkage based high sensitivity large band monolithic seismometers and accelerometers for geophysics and seismology Proc. of 16th WCEE* Santiago, Chile.

[10] Barone F, Giordano G, Acernese F, Romano R, Gennai A, Passuello D, Boschi V, Cerretani G, Passaquieti R 2016 *Large band low frequency sensors based on Watts linkage for future generations of interferometric detectors GWADW 2016* Isola d-Elba, Italy.

[11] Wu S, Fan S and Luo J, 2002 *Folded pendulum tiltmeter Rev. Sci. Instrum.* **73** 2150 doi:10.1063/1.1469676.

[12] Takamori A, Bertolini A, DeSalvo R, Araya A, Kanazawa T, Shinohara M 2011 *Novel compact tiltmeter for ocean bottom and other frontier observations Meas. Sci. Technol.* **22** 1115901 doi:10.1088/0957-0233/22/11/115901.

[13] Liu J F, Ju L, Blair D G 1997 *Vibration isolation performance of an ultra low frequency folded pendulum resonator Phys. Lett. A* **228** 243 doi:10.1016/S0375-9601(97)00105-9.

[14] Bertolini A 2001 *High Sensitivity Accelerometers for Gravity Experiments* Ph.D Thesis, LIGO P0100009-00-Z.

[15] Barone F, Giordano G, Acernese F, *Method for the measurement of angular and/or linear displacements utilizing one or more folded pendula* Pat. Num.: IT1425605, Europe, Japan, USA, Canada pending.

IOP Conf. Series: Journal of Physics: Conf. Series **957** (2018) 012012 doi:10.1088/1742-6596/957/1/012012

Status of the Monolithic Suspensions for Advanced Virgo

F Travasso[1,2] on behalf of Virgo Collaboration
[1]EGO, via Amaldi, S. Stefano a Macerata - Cascina (PI) 56021, IT
[2]INFN - Perugia, via Pascoli, Perugia 06125, IT

E-mail: `flavio.travasso@pg.infn.it`

Abstract.
Successfully implemented in GEO and Virgo+, the monolithic suspensions are one of the most important upgrades in the second generation of gravitational wave interferometric detectors, including Advanced LIGO (aLIGO) and Advanced Virgo (AdV). Characterized by a very low thermal noise, monolithic suspensions are essential for improving the interferometers sensitivity at low frequencies (10-100Hz). In Advanced Virgo their installation was delayed because of a contamination problem in the vacuum system: dust produced by scroll pumps was injected in the main vacuum chambers during the venting processes, damaging the fibers and ultimately causing their repeated failure. The effort to explain and resolve this issue was useful to further confirm the suspensions' reliability and our control on the production process. Moreover, we developed and implemented new tools and procedures to certify each part of the monolithic suspensions. In the meanwhile, in order to join aLIGO during its second Observation Run (O2), a temporary steel suspension was implemented, based on the initial Virgo design. That solution allowed us to contribute to the first three-detector observation of a gravitational wave (GW) ([1]), and to the first observation of a coalescing neutron star binary ([2]) In the near future the monolithic suspensions will be reinstalled along with additional upgrades of Virgo.

1. Introduction

In AdV, the input and end mirrors are all suspended from a steel *marionette* which, in turn, is suspended to the last filter of the superattenuator,called *filter7*. The marionette and the mirror form a double-pendulum system, surrounded by a rigid structure called *cage*, attached to the *filter7* [3]. The assembly of the marionette-mirror pendulum system with the cage and associated auxiliary optics and actuators is referred to as the *payload*. The payload is conceived for the compensation of residual seismic motion and for acting on the test masses working point in the interferometer. A schematic view of AdV payload is shown in Fig.4, whereas a full description is available in the paper [4]. In this final stage two silica interface pieces (ears) are bonded to the mirror using the silicate bonding technique [5]. Four silica fibers are glued under those ears and on the top of the marionette with the same technique (Fig.1), providing better seismic isolation and lower thermal noise [6, 7]. In the following, we shall deal mainly with the development and implementation of our monolithic suspensions in AdV.

Content from this work may be used under the terms of the Creative Commons Attribution 3.0 licence. Any further distribution of this work must maintain attribution to the author(s) and the title of the work, journal citation and DOI.
Published under licence by IOP Publishing Ltd

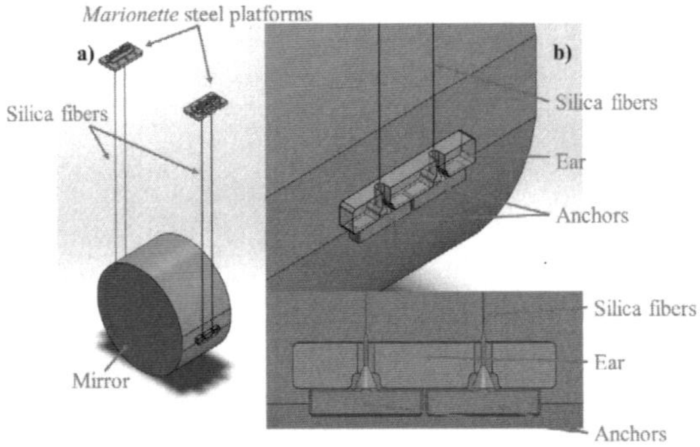

Figure 1. Monolithic suspension design: a) Schematic mirror suspension; b) Details of anchors and ear.

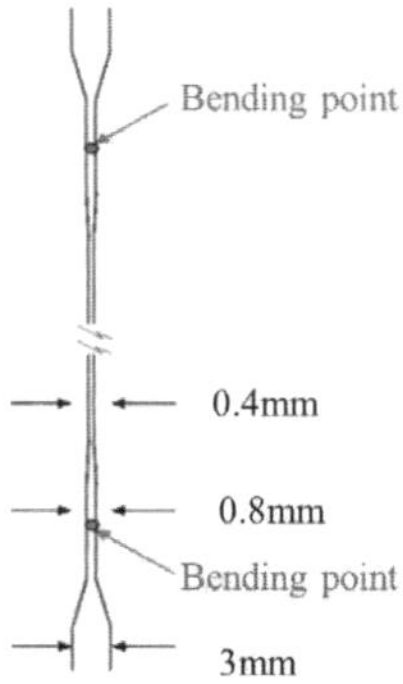

Figure 2. Fiber profile.

2. Advanced Virgo Monolithic suspensions

The monolithic suspensions were developed keeping into account the mechanical constrains of the whole payload, with the aim of reducing the suspension thermal noise and increasing the reliability of the suspensions themselves [6, 3]. The installation procedure was defined by performing dozens of simulations and tests on each part of the suspension system [3]. Moreover a recovery strategy was developed and tested using dummy parts: a brief description is available in the *Recovery strategy* paragraph of present paper.

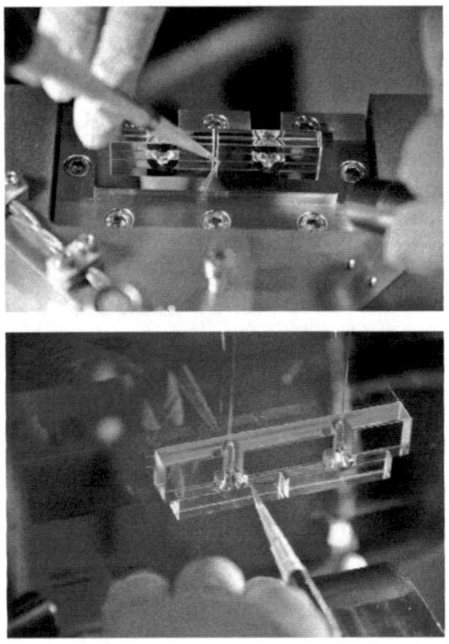

Figure 3. Gluing process.

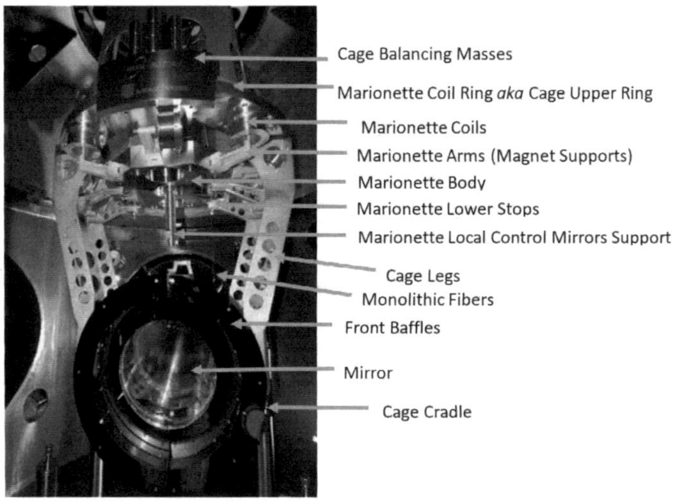

Figure 4. AdV payload [4].

2.1. Fibers design and production

The test masses, having a mass of 42 kg, are suspended from the marionette with the monolithic suspension system. The design of the monolithic suspensions is optimised taking into account the issues related to the thermal noise reduction, the mechanical frequencies and the stresses in the silica elements as well as the mounting, handling and recovery procedures. An extensive finite element simulation was used for this purpose (see [8] for the anchors and ears).

The fiber profile is shown in figure 2; this dumbbell shaped geometry strongly reduces the thermoelastic contribution to the thermal noise [9]. Fibers are produced starting from a rod, welded to fused silica anchors (Fig.5), using a CO_2 laser pulling machine [3, 10]. All the fiber parameters are carefully measured during this process, such as the fiber length under nominal load, in order to meet the requirements on mirror positioning and the bending length [11, 12]. Statistics on about 100 fiber samples measured a breaking load of about 40 kg ($\sim$ 3 GPa), i.e. four times the working load. Validation of each fiber is done applying briefly a load of 20 kg ($\sim$ 1.5 GPa) load.

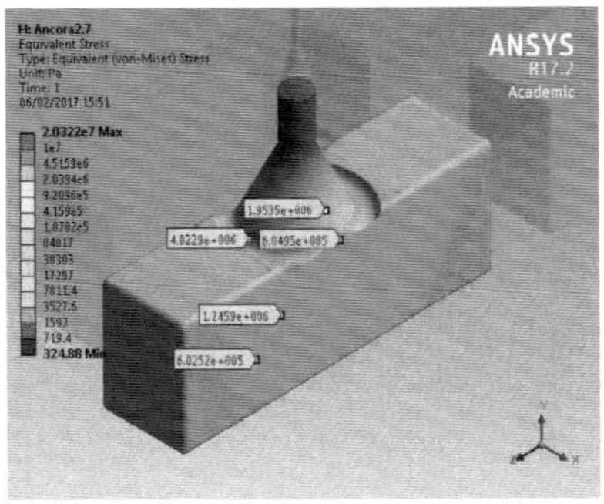

Figure 5. Dissipation and stress simulation to optimize the silica parts and to reduce their thermal noise [8].

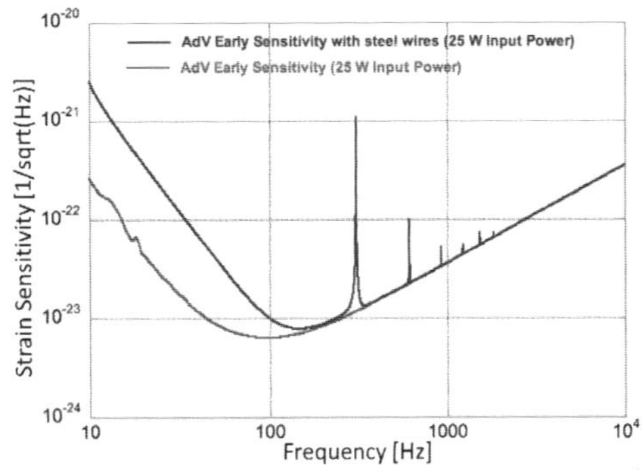

Figure 6. AdV sensitivity curve.

3. Monolithic issue

Successfully implemented in Virgo+ [10], the monolithic suspensions failed in AdV during the installation phase. A summary of the breakages is shown in Tab.1

Mirror failed	Failure date	Time in air	Time in vacuum
WI	18 November 2015	5 months	7 days
NI	18 Dicember 2015	4.5 months	5 days
NI	1 March 2016	1 week	5 days
WI dummy	26 April 2016	1 week	11 days
WI	25 June 2016	1 month	1 month
WE	28 June 2016	7 months	18 days
NE	13 October 2016	6 months	4 months

Table 1. Advanced Virgo monolithic suspensions failures summary

After the first failure, several possible reasons were analysed in laboratories of Virgo and of the LIGO Scientific Collaboration [13]. All the analyses didn't show any significant evidence of a problem, however there were few indications that the breakages of the fibers were not the first cause of the suspension failures but just the result of something else.

On October 13th, 2016, the NE payload failed during a venting operation. It had been in air for 6 months and in vacuum for 4 months (Tab.1). Starting from this event it was finally realized that there was a contamination problem in the pumping system. This hypothesis was confirmed by the fact that the failures always started from the fibers closer to the venting side. Moreover, a report by an external company (SSV laboratory), issued just at the time of the NE event, confirmed that all the failures had started from the thin part of the fibers: it became evident that something from the venting pipe was hitting the fibers.

4. Vacuum issue

After the failure of the NE payload, clearly related to a venting operation, two facilities were developed to test both a single fiber and an entire payload. The two setups replicated as much as possible the standard vacuum system of a Virgo vacuum chamber (the *tower*) and were used to test the venting procedure. With the single fiber setup was possible to reproduce, for the first time, a failure a few seconds after the venting of the chamber. The test was repeated three times with the same result, but with different failure times. The dust collected from these tests and from the towers after the failures was analyzed by optical microscope, SEM, IR and Raman spectroscopy. All these tests confirmed that the contamination was due to particles coming out from the scroll pumps. Indeed, in the standard vacuum configuration the scroll pumps and the venting system shared the same pipe: the failures were due to the dust produced by the scroll pumps, stored in the pipe and then injected in the main vacuum chamber during the venting process, hitting the fibers.

An analytical model developed by the collaboration [14] shows that, during a venting, air shock waves can accelerate 50 - 100 micron particles to speeds up to 20-50 m/s for a 2 m travel (Fig.7). The kinetic energy of these particles is well in the range necessary to cause the breaking of a fiber.

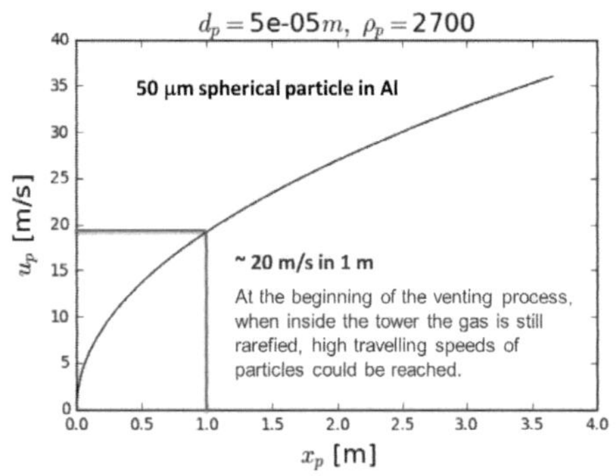

Figure 7. Analytical model of the speed of the dust in vacuum.

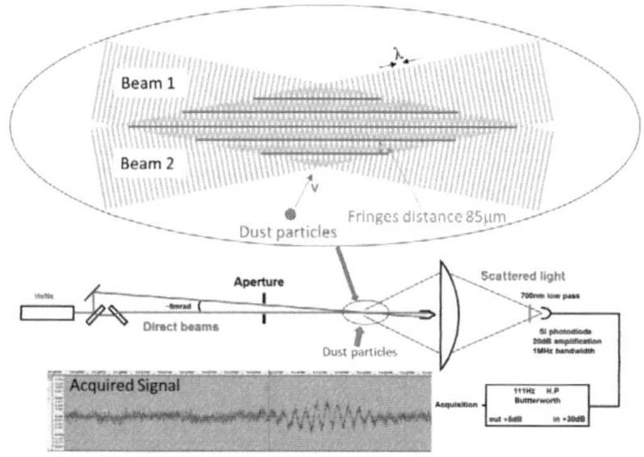

Figure 8. Velocimeter optical scheme

To verify this hypothesis, an interferometric velocimeter was developed and used to measure the velocities of particles coming out of a scroll pump (Fig.8): velocities of the order of 1 m/s were observed. An interesting addendum was proposed by the study of the effects of an electrostatic field: it increases the effective cross section of the fibers for incoming particles; on charged surfaces the dust is 5 times more w.r.t. the not charged ones.

It is worth to remark that during the Virgo+ runs, the vacuum contamination problems were probably avoided because pots were in place that covered about half of the section of the tubes used for venting and pumping. Moreover, in Virgo+ scroll pumps were used with a reduced duty cycle.

After the identification of this problem, new features were discussed with external experts of clean vacuum applications from CERN, LASA and DESY. The main upgrades that will be implemented concern the use of new *clean* pumps with longer pipes and particle filters; the design of a separated venting circuit; the installation of an inner shield in front of the venting pipe to stop any residual contaminants; the use of fiber guards on the payload itself [4]. Moreover, we shall perform a special cleaning procedures of all tower chambers to remove any residuals from the past ventings with scroll pumps and from the explosion of the silica fibers, before each new integration of a monolithic payload.

5. Temporary solution for O2: steel suspensions

To proceed with the commissioning of the detector, the Virgo Collaboration asked for a temporary solution that allowed to test the different sub-systems in order to join O2 in coincidence with LIGO. The faster and more reliable solution was the implementation of steel suspensions such as those installed during the first Virgo setup [15]. To suspend the main optics, two steel wires were used, forming a cradle on which the mirror was to be housed. The wires, with a diameter of 0.3 mm, were thermally treated for one week at 150 °C with their nominal load, to release their internal stress and to reduce the creep noise. To connect the wires to the steering system, four steel clamps were screwed on the marionette while four silica stand-offs were glued on the lateral flats of the mirrors, in order to reduce the friction between the wires and the mirror barrel.

The expected sensitivity curve in this configuration is shown in Fig.6 [16]. This solution made possible the first three-detectors observation of gravitational waves from a binary black hole coalescence (GW170814) [1] and the first multi-messenger observation of a binary neutron star merger (GW170817) [2].

6. Monolithic suspensions upgrades

To improve the AdV sensitivity, the monolithic suspensions will be reinstalled for the O3 observation run in coincidence with LIGO interferometers (Fig.6). Although all the analyses have shown that there was nothing wrong with the current design of monolithic suspensions, the knowledge acquired during the revision process was useful to confirm the suspensions reliability. In addition, an optimized procedure for fiber production was developed, a new anchor design was devised, new tools to certify each part of the suspensions were implemented, a better recovery and reinstallation strategy was defined.

6.1. Fiber upgrades

Even though there was no evidence of any weakness in the fiber production process, the whole procedure has been revised and optimized. The welding is now obtained with a longer laser exposure time, at lower power according to Glasgow group suggestion. A digital camera ensures a good control of the welding process, and an optical scan system with a digital microscope is used to search for any defect. Furthermore, a new drawing for the anchors was implemented using a thicker T-shape design with a smoother groove around the cone. The anchors were treated with a flame polishing and successive annealing to reduce the cracks induced by the machining. With this new design and procedures, two breaking stress tests were performed and the fiber/anchors system reached respectively 160kg and 130kg before breaking, 10 times greater than the nominal load. The tests were recorded by a high speed camera (about 15000 fps) to identify the origin and the dynamics of the breakages.

Furthermore, in the test payload facility at the Virgo site some studies are in progress, in order to reduce the suspension thermal noise. In particular the use of a new generation shadow-meter [17] will be very useful to study the dynamics of the fibers thermal peaks and the possibility to damp them.

6.2. Recovery strategy

The payload will be removed from the tower, placed in a clean room and dismounted, and the mirror will be put in its transportation box and stored under a class 1 clean hood. According to the past experiences, to clean the mirror it is sufficient to apply the *first contact*; as usual, it will be peeled away just before closing the flange of the tower for pumping down its vacuum chamber. After the application of the *first contact*, it will be possible to remove safely the spacers, used for the steel wire suspensions, and to clean the ears from any residual of the anchors, left from the suspension failures. Holes and scratches due to the failures will not be cured: indeed our FEM simulations show that if the thickness of the silicate bonding is greater than 8.5 μm, its thermal noise starts to dominate on the overall thermal noise of the mirrors. Hence, the recommendation is to clean the ear from any residuals and to avoid any glue in the holes.

At the end of this procedure, the mirror will be ready to be suspended in the payload. The complete reinstallation procedure will last three week per payload. Before starting the real reassembling of the monolithic payloads, a full scale test will be performed by integrating the test payload, with real fibers and a dummy mirror, inside the NE tower. The dummy payload will be equipped with the new fiber guards and with supports holding wafers for contamination control [4]: a dedicate optical scan for wafer analysis was developed to certify the vacuum quality. A full description of all the procedures is available in [18].

7. Conclusion

Successfully implemented in Virgo+, the installation of our monolithic suspensions in Advanced Virgo was delayed because of a series of failures that happened under vacuum after days or months from the integration of the payloads. These problems forced us to suspend the test mass mirrors with steel wires, so to be able to run in coincidence with aLIGO in O2, while, at the same time, performing tests and studies on the failures. After about one year of investigations, finally we found that some particles, coming from the pumping system, were injected into the towers during the venting cycles, and projected onto the fibers. The amount, size and kind of particles varied in an unpredictable way over time and from one tower to the other, resulting in different kinds of damages to the fibers and, consequently, in different times of failure. This scenario is confirmed by all tests, analysis and simulations carried out up to now.

The main message coming from these results is that there was nothing wrong with the current design of our monolithic suspensions. In any case, the many investigations carried out since the first suspension failure in November 2015 resulted in a careful revision of all our procedures and in further refinements of the design and testing procedures, confirming the reliability of our monolithic suspensions and improving our control on the production process.

A positive results of this long quest for finding a solution to the monolithic suspensions failures was the development of closer collaborations with other monolithic suspensions group in the LIGO Scientific Collaboration. In particular, we have strengthened our collaboration with the Glasgow Group, whose help has been precious during these investigations.

By the end of this year (2017), we shall start to integrate again monolithic suspension in Advanced Virgo so to join aLIGO with an enhanced sensitivity in O3.

References

[1] Abbott B P *et al.* 2017 *Phys. Rev. Lett.* **119** 141101
[2] Abbott B P *et al.* 2017 *Phys. Rev. Lett.* **119** 161101
[3] Acernese F *et al.* 2015 *Class. Quantum Grav.* **32** 024001
[4] Naticchioni L, in preparation to be published in this volume of *Journal of Physics: Conference Series*
[5] Gwo D H 1998 *Proc. SPIE* **3435** 136-142; see also US Patent No. 6,548,176 (1998)
[6] Amico P *et al.* 2002 *Rev. Sci. Instrum.* **73** 3318-3323
[7] Amico P *et al.* 2004 *Nucl. Instrum. Methods - A* **518** 240-243
[8] Puppo P 2017 *Virgo report* VIR-0494A-17 (https://tds.virgo-gw.eu/ql/?c=12425)
[9] Cagnoli G and Willems PA 2002 *Phys. Rev. B* **65** 174111
[10] Aisa D *et al.* 2016 *Nucl. Instrum. Methods Phys. Res. Sect. A-Acc. Spectrum. Dect. Assoc. Equip.* **824** 644
[11] Lorenzini M *et al.* 2013 *Rev. Sci. Instrum.* **84** 033904
[12] Piergiovanni F *et al.* 2010 *J. Phys. Conf. Ser.* **228** 012017
[13] Travasso F 2016 *Virgo Report* VIR-0383A-16 (https://tds.virgo-gw.eu/?content=3&r=13723)
[14] Viceré A 2016 *Virgo Report* VIR-0516A-16 (https://tds.virgo-gw.eu/?content=3&r=12741)
[15] Cagnoli G, Gammaitoni L, Kovalik J, Marchesoni F and Punturo M 2000 *Rev. Sci. Instrum.* **71** 2206-2010
[16] Guidi G and Puppo P 2016 *Virgo report* VIR-0316F-16 (https://tds.virgo-gw.eu/ql/?c=11627)
[17] Lockerbie N A and Tokmakov K V 2014 *Meas. Sci. Technol.* **25** 125110
[18] PAY team 2017 *Virgo Report* VIR-0712A-17 (https://tds.virgo-gw.eu/ql/?c=12645)

IOP Conf. Series: Journal of Physics: Conf. Series **957** (2018) 012013 doi:10.1088/1742-6596/957/1/012013

Schenberg microwave cabling seismic isolation.

Bortoli, F.S.[1] ,Frajuca, C.[1]and Aguiar, O.D.[2]
[1] Sao Paulo Federal Institute – Sao Paulo, SP, Brazil

[2] INPE – Divisão de Astrofísica, São Jose dos Campos, SP, Brazil

E-mail: Frajuca@gmail.com

Abstract: SCHENBERG is a resonant-mass gravitational wave detector with a frequency about 3.2 kHz. Its spherical antenna, weighing 1.15 metric ton, is connected to the external world by a system which must attenuate seismic noise. When a gravitational wave passes the antenna vibrates, its motion is monitored by transducers. These parametric transducers uses microwaves carried by coaxial cables that are also connected to the external world, they also carry seismic noise. In this analysis the system was modeled using finite element method. This work shows that the addition of masses along these cables can decrease this noise, so that this noise is below the thermal noise of the detector when operating at 50 mK.

1. Introduction

The detection of Gravitational Waves (OG) came after a long road of efforts [1,2,3,4,5,6,7,8,9]. The Brazilian efforts towards the detection of gravitational waves are centered on the Schenberg detector. In the Schenberg detector six sensors are connected to the surface of the sphere, arranged according to the distribution of Merkowitz and Johnson [10,11]. These transducers are located as if they were in the center of 6 pentagons connected in a surface corresponding to half dodecahedron. Each transducer mechanically amplifies the motion occurring on the region of the sphere in which it is connected. The already amplified movement excites the membrane of one resonant cavity. In this resonant cavity microwaves are pumped, which generate the electronic signal that will return taking all the information of the OG's. Intensity and direction of the OGs can be obtained from the analysis of the output signal of these 6 transducers.

To reach the resonant cavities, first the microwaves are conducted from the outside of the dewar (thermo flask where every antenna system is contained) by cabling to microstrip antennas. These antennas, located in front of the parametric transducers, conduct the microwaves into the resonant cavity and another set of antennas pick up the returned signal. The Brazilian efforts on the field can be seen in [12,14,15,16,17,18,19,20, 21,22,23,24,25,26,27]

The cabling has two parts: the first comes from the exterior of the dewar and mechanically connects the last mass of the sphete suspension (lower mass, will be refered here as mass 5) and the second, which is also connected to this mass, continues until it connects to the microstrip antennas. Here the analysis will be done on the first section of the cabling and will propose a way to reduce the seismic noise introduced by the cabling, so that the noise on the transducers does not become greater than the thermal noise. This analysis, as well as the others, were done using simulation using the Finite Element Method (FEM) [28,29].

2. Necessary noise attenuation

Taking into account Schenber detector parameters the displacement noise square by thermal noise is estimated at 10^{-46} m^2/Hz (this dimension is chosen because it is easier to work in FEM). The detector site parameters for seismic noise was estimated at 10^{-30} m^2/Hz. Under a pessimistic approach this value was used to excite one end of the cabling, more specifically, the one connected to the environment outside the detector. Thus, from the point of the cabling in contact with the external environment to the region of the sphere where the transducers are connected, it is necessary to obtain an attenuation of the order of 10^{16} or 16 decades.

Content from this work may be used under the terms of the Creative Commons Attribution 3.0 licence. Any further distribution of this work must maintain attribution to the author(s) and the title of the work, journal citation and DOI.
Published under licence by IOP Publishing Ltd

IOP Conf. Series: Journal of Physics: Conf. Series **957** (2018) 012013 doi:10.1088/1742-6596/957/1/012013

3. The proposed attenuation system

To obtain the required attenuation, the use of a mass-spring system using the cabling itself was studied. In each cable, a plurality of masses were applied in the form of small disks and spaced apart in such a way that each portion of the cable functions as a spring of this mass-spring system. If this system is properly calibrated, the movement due to the seismic noise that excites the cable will be partially reflected by each of these masses, so that the total attenuation produced by all the masses is sufficient so that there is no interference in the measurements made by the detector .

4. Selecting the best cabling configuration using FEM

Several simulations were performed using the finite element method, in which several configurations were tested. In a first step a simpler, symmetrical and constant radius cable was used. The more complex systems were testes, in each simulation the thicknesses of each disk (here used only as mass in a mass-spring system) and the spacing between them along the cabling were changed. Figures 1 and 2 show one of these simulated configurations and its graph of frequency response, where URES is the resultant displacement squared.

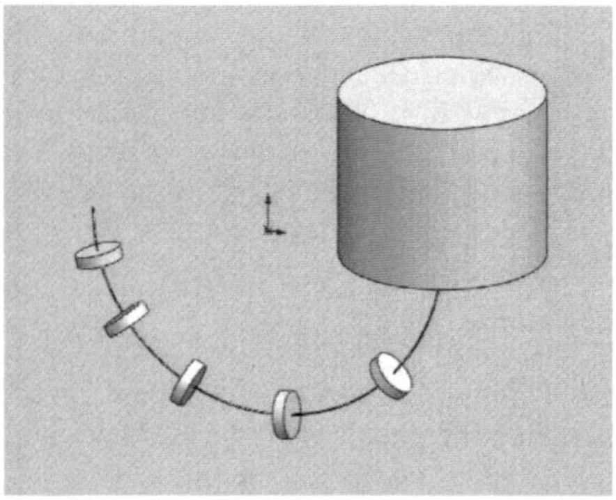

Figure 1: System composed of the copper cable connected to five disks with a diameter of 20mm and a thickness of 5mm and the rigid mass (rigid mass meaning the this mas has no vibration modes).

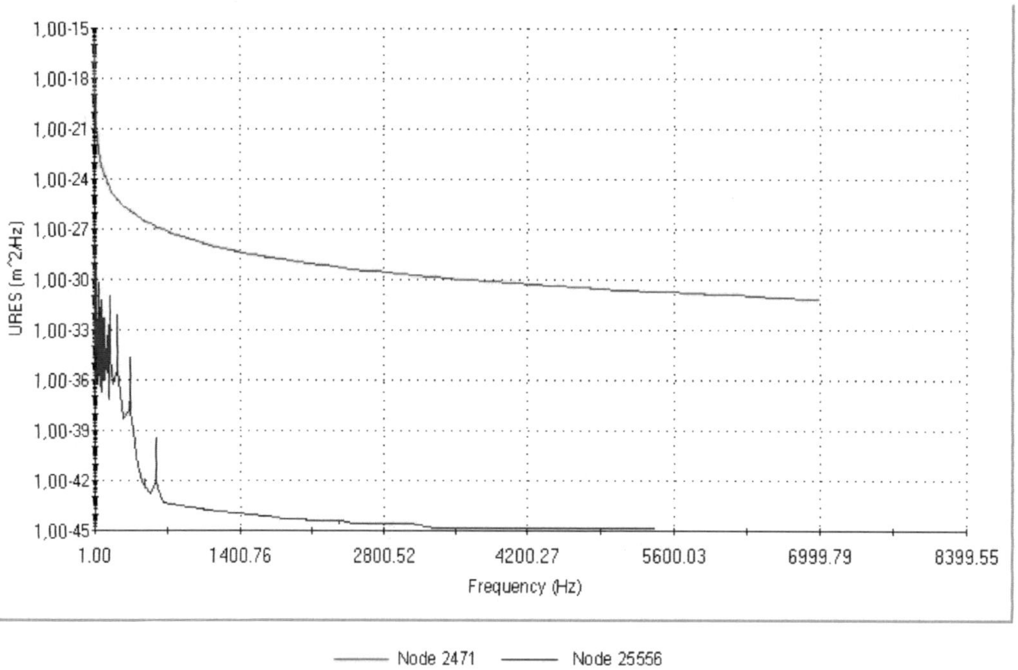

Figure 2: Frequency response curves for the system shown in Figure 1. The upper curve refers to the point at the excited end (left side point) and the other to the point at the end where the rigid mass is connected.

IOP Conf. Series: Journal of Physics: Conf. Series **957** (2018) 012013 doi:10.1088/1742-6596/957/1/012013

5 Some conclusions obtained up to this stage

The main characteristics observed when analyzing the response curves were the attenuation produced in each configuration used and the quality of this attenuation.

In the graphs obtained the attenuation produced in a given frequency is the measure of the distance between the curve refering to the place where the excitation occurred and the curve of the analyzed point. This measure is given in m^2/Hz by the program.

The quality of the attenuation is observed from the oscillation of the curve in the frequencies of interest.

Taking into account these criteria, the analysis of the Frequency Response Curves obtained in this stage led to the following conclusions: the addition of equal disks increases the attenuation, but this is only representative up to the fifth mass added; the quality of the attenuated curve is improved by the addition of equal disks; increasing the size of all masses connected to the cable does not produce a significant increase in attenuation, although slightly improving the quality of the attenuated curve and the increase of the radius of curvature improves the attenuation, and the best value for the attenuation with this configuration (10^{18}) was obtained, although the attenuated curve did not present good quality.

The attenuation obtained with a cable with 80mm length and another with a length of 160mm, both having 5 disks (masses), with decreasing spaces, demonstrated this but there no space to shown it here. With the longer cable an attenuation of the order of 10^{18} was achieved around the frequency of 2530 Hz, whereas with the shorter length this attenuation was of the order of 10^{15}. As there is space constraint inside the dewar, it is not possible to freely increase the cable length to improve the attenuation. The use of disks with different thicknesses did not alter the attenuation or the quality of the attenuated curve. The use of uneven cabling showed a slight improvement in attenuation, this can be attributed to the longer cable length in this case but is nor practical.

It is important to note that the peaks observed in the frequencies of interest could be shifted to other regions by appropriate calibration of the parameters involved, thus altering the "windows" around frequencies of interest. The results for the attenuation obtained with the models tested so far was 10^{15} to 10^{18}.

5.1 Cable with uneven ends, remote mass and five disks

With the knowledge of the systems studied so far a realistic configuration was designed. That is, one that takes into account the space available inside the Dewar, as well as the position of the ends of the cabling (which depend on the position of their connections), now better defined by the overall design. In addition to defining the horizontal and vertical distances between the connectors at the ends of the cabling, the overall cable length has also been defined. In this way, a project that is adequate to the detector is made, from the study of a cabling that is appropriate to the constraints and real characteristics of the project.

Figure 3 shows the best configuration tested, as well as having a format suitable for the detector design. In this configuration the ends of the cable are in 190mm difference in height, have a vertical projection distance of 190mm and the total cable length is 460mm. The disks are positioned in such a way that they have increasing distances and their masses are decreasing.

Table 1 shows the diameters of the disks, their spacings (measured from the end of the cable that was excited), their thicknesses and their masses.

Figure 4 shows the frequency response graph of this configuration, where the attenuation of 10^{16} is shown for the 3200Hz and vicinity frequency.

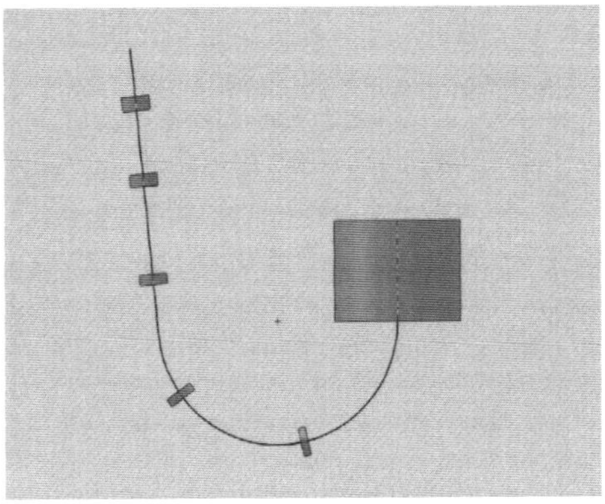

Figure 3: System composed of the remote mass, the copper cable with a diameter of 0.5 mm and a total length of 490 mm, a horizontal spacing of 190 mm and a 190 mm difference between the ends of the cable. The five disks with a diameter of 20mm and thicknesses of 10.0 - 9.0 - 7.5 - 6.0 – 5.0 mm were connected to the cable with increasing distances between the disks. The excitation considered acts on the left end of the cable.

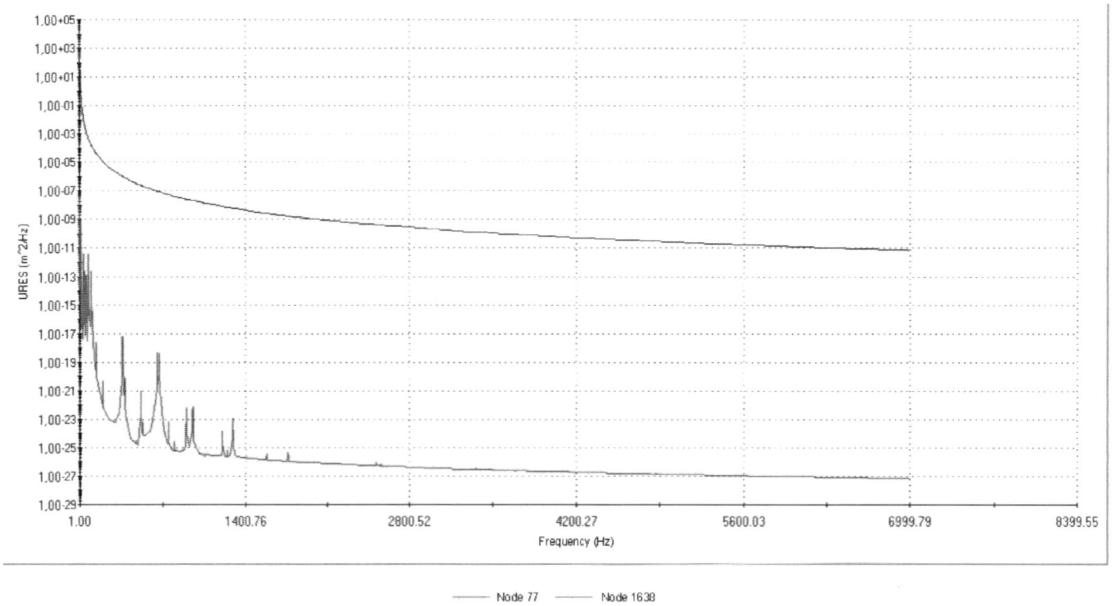

Figure 4: Frequency response curve for the system shown in figure 3 obtained for two points of the cabling. One point is situated at the excited (left) end and the other at the end where the rigid mass is connected.

Table 1: Dimensions and masses of the disks of the model of figure 3. The table shows the diameters of the disks, the distances between them (measured from the end of the cable that was excited), their thicknesses and their masses.

	Diameter (mm)	Gap (mm)	Thickness (mm)	Mass (g)
1	20	38.3	10.0	28.2
2	20	92.0	9.0	25.4
3	20	161.0	7.5	21.1
4	20	245.3	6.0	16.9
5	20	345.0	5.0	14.1

6 Frequency response of the seismic noise originated in the cabling at the transducer locations on the sphere

After some conclusions a cable configuration with a more realistic conformation and dimensions compatible with the available space within the Dewar was used.

Having obtained a good cabling design, a simulation was performed to evaluate the remaining seismic noise and the attenuation, on the surface of the sphere, in the same places where the transducers will be connected.

Figure 5 shows in detail the cabling with the best configuration obtained, connected to suspension last mass 5 (located at the bottom of the suspension) and an overview of the suspension with cabling. To simplify the simulation, the model does not include connectors and other elements, but this should not compromise the results, as these elements are rigidly attached to the suspension.

The effect produced by the seismic noise on the locations where the transducers will be connected, obtained in the simulation, is shown in figure 6. In this figure the locations of the transducers are identified as upper and lower cavities. In the simulation the excitation was made only through the free end of the cabling in the model. It is estimated that the addition of the other cables would not result in a significant increase in noise. In the response graph of this analysis it can be observed that: an attenuation of the order of 20 decades (10^{20}) between the excited end of the cable and the underside of the mass 5 and an additional 19 decades attenuation between the lower face of the mass 5 and the upper and lower cavities on the surface of the sphere.

Thus, the attenuation between the excited end of the cabling and the transducer locations was estimated at 39 decades at frequencies of the order of 3200 Hz.

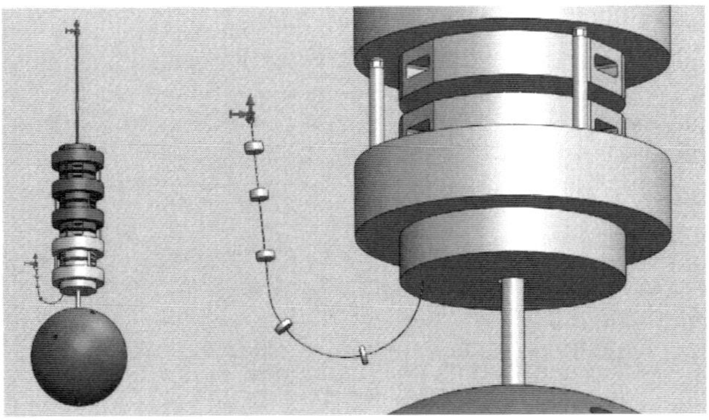

Figure 5: Cabling model using the best mass-cable attenuator (mass 5 is located at the bottom of the suspension) and an overview of the suspension with cabling.

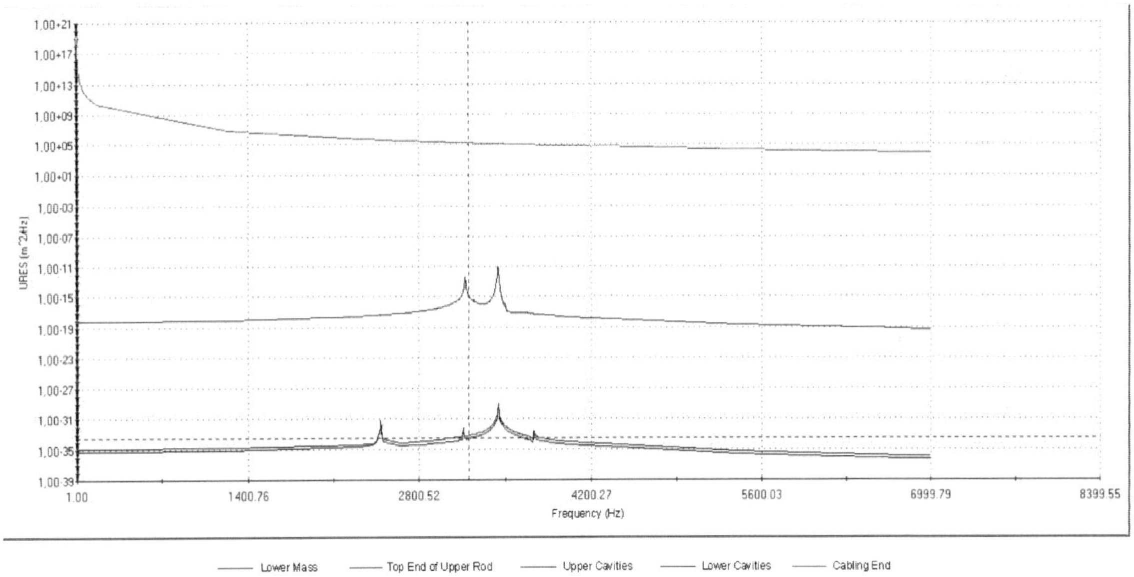

Figure 6: Frequency Response Curve for cabling model connected to suspension mass 5 (mass located at the bottom of the suspension). In this analysis, the effect of seismic noise from the end of the cabling on the locations where the transducers are connected, identified here as upper and lower cavities, were evaluated in this analysis. The excitation was made through the cabling end and can be seen in the upper curve.

7 Conclusions

The effect produced by the seismic noise that is transported by the cabling, on the places where the transducers will be connected, was obtained in the simulation. In the model the excitation was made only at the end of the cabling, to consider only the effect of the cables. It is estimated that the addition of the other cables will not result in a significant increase in noise. In the response graph of this analysis it can be observed that: an attenuation of the order of 20 decades (10^{20}) between the excited end of the cable and the the lower face of mass 5; and an additional 19 decades attenuation between the lower face of the mass 5 and the cavities upper and lower surfaces of the sphere.

Thus the attenuation between the excited end of the cabling and the transducer locations has been estimated at 39 decades at frequencies of the order of 3200Hz, attenuation is sufficient so that the sensitivity designed for the detector can be achieved.

Acknowledgements
Carlos Frajuca acknowledgs FAPESP for grant #2013/26258-4 and grant #2006/56041-3.

References

[1] The Gravitational Waves International Committee Roadmap (GWIC). A global pan. June 2010. Glasgow: Univerty of Glasgow - Department of Physics and Astronomy - Kelvin Building (G12 8QQ), 117p.
[2] Taylor J H, Hulse R A, Fowler L A, Gullahorn G E, Rankin J M 1976 *Astrophysical Journal* **206** L53
[3] Weber J. 1960 *Physical Review* **117** 306
[4] Thorne K S 1987 "300 years of gravitation". Cambridge: Cambridge University Press: 1987, p.330.
[5] Blair D G "The detection of Gravitational Waves." 1991 Cambridge: Cambridge University Press
[6] Richard J P 1984 *Physical Review Letters* **167** 165
[7] Richard J P Proc. Second Marcel Grossmann Meeting on General Relativity, edited R.Ruffini, (North-Holland, Amsterdam).
[8] Frajuca C *et al* 2004 *Class. Quantum Grav.* **21** 1107
[9] G. Frossati, Proc. First International Workshop for an Omnidirectional Gravitational Radiation Observatory, W.F. Velloso, Jr., O.D. Aguiar and N.S. Magalhaes, editors (Singapore, World Scientific, 1997).
[10] Merkowitz S M and Johnson W W 1997 *Phys. Rev. D* **56** 7513
[11] S. M. Merkowitz and W. W. Johnson 1993 P*hys. Rev. Lett.* **70**, 2367
[12] Frajuca, C. "Otimização de transdutores de dois modos mecânicos para detectores de ondas gravitacionais." São Paulo. 97 p. Phd Dissertation - Universidade de São Paulo, 1996.
[13] Frajuca C. *et al* , Proc. 3rd Edoardo Amaldi Conference on Gravitational Waves (Pasadena, USA, July 1999). AIP Conf. Proc. 523 (New York, AIP) p.417
[14] Aguiar O D et all 2011 *Journal of Physics: Conference Series* **363** 012003
[15] O. D. Aguiar et. al 2005 *Class. Quantum Grav.* **22**, 209
[16] Frajuca *et al* 2002 *Class. Quantum Grav.* **19** 1961
[17] Magalhaes N S *et al* 1997 *Astrophysical Journal* **475**, 462
[18] Magalhaes N S *et al* 1995 *MNRAS* **274**, 670
[19] Frajuca C, Bortoli F S, Magalhães N S 2005 *Brazilian Journal of Physics* **35** 1201
[20] Aguiar O D et al. 2006 *Journal Class. Quantum Grav.* **23**, 239
[21] Frajuca C, Bortoli F S, Magalhaes N S 2006 Journal of Physics: Conference Series **32** 319
[22] Frajuca C, Bortoli F S, Magalhães N S 2005 *Brazilian Journal of Physics* **35** 1201
[23] Aguiar O D et al., *Class. Quantum Grav.* **21** 459, 2004.
[24] Frajuca C, Magalhaes N S, Horiguti A M 2008 *Journal of Physics: Conference Series* **122** 012029
[25] Bortoli F S *et al* 2010 *Journal of Physics: Conference Series* **228** 012011.
[26] Magalhaes *et al* 1997 *Gen. Relat. Grav.* **29** 1511
[27] Magalhaes *et al* 1997 *Astroph. J.* **475** 462
[28] Bathe, Klaus-Jürgen, "Finite Element Procedures", New Jersey: Prentice Hall, 1996.
[29] Filho A A, "Elementos Finitos", Sao Paulo: Erica, 2000

IOP Conf. Series: Journal of Physics: Conf. Series **957** (2018) 012014 doi:10.1088/1742-6596/957/1/012014

Surprises from the spins: astrophysics and relativity with detections of spinning black-hole mergers

Davide Gerosa[1]

TAPIR 350-17, California Institute of Technology,
1200 E California Boulevard, Pasadena, CA 91125, USA

E-mail: dgerosa@caltech.edu

Abstract. Measurements of black-hole spins are of crucial importance to fulfill the promise of gravitational-wave astronomy. On the astrophysics side, spins are perhaps the cleanest indicator of black-hole evolutionary processes, thus providing a preferred way to discriminate how LIGO's black holes form. On the relativity side, spins are responsible for peculiar dynamical phenomena (from precessional modulations in the long inspiral to gravitational-wave recoils at merger) which encode precious information on the underlying astrophysical processes. I present some examples to explore this deep and fascinating interplay between spin dynamics (relativity) and environmental effects (astrophysics). Black-hole spins indeed hide remarkable surprises on both fronts: morphologies, resonances, constraints on supernova kicks, multiple merger generations and more...

These findings were presented at 12$^{\rm th}$ Edoardo Amaldi Conference on Gravitational Waves, held on July 9-14, 2017 in Pasadena, CA, USA.

1. Introduction

The second-generation gravitational-wave (GW) detectors Advanced LIGO and Virgo are taking us into the age of observational strong-field gravity. The detections of black-hole (BH) binaries has now become routine with 4.87 events observed during Advanced LIGO/Virgo first and second observing runs (4 confirmed detections [1, 2, 3, 4] and 1 candidates with 87% probability of being of astrophysical origin [5]). These observations are giving us unique insights on a sector of the Universe we hardly knew existed. Three of the detected binaries (GW150914, GW170104 and GW170814 [1, 3, 4] have component masses $\sim 30 M_\odot$, largest than all stellar-mass BHs known from X-ray binary observations [6]. The event GW151226 [2] has a lower mass, but a confirmed detection of BH spin.

Despite the low statistics, these first BH detections are already providing invaluable information on the life of their stellar progenitors. For instance, the unexpected high mass of some of the events can only be modeled if systems form in a low metallicity environment ($Z \lesssim 0.5 Z_\odot$), where stellar winds are suppressed [7, 8]. More observations will become available in the next few years as detectors keep on being improved, thus allowing us to enter the large-statistics regime and measure details of the underlying astrophysical populations of BHs in binaries.

[1] Einstein Fellow

Content from this work may be used under the terms of the Creative Commons Attribution 3.0 licence. Any further distribution of this work must maintain attribution to the author(s) and the title of the work, journal citation and DOI.
Published under licence by IOP Publishing Ltd

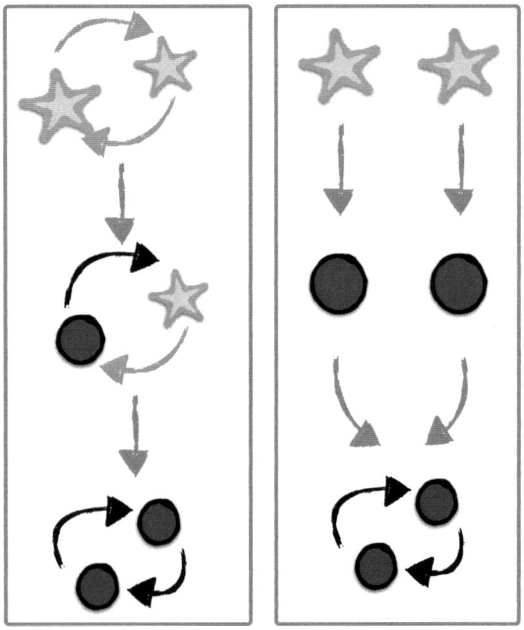

Figure 1. Two main classes of stellar-based BH formation models. BH binaries can form *in the field* from binary stars (left): both stars collapse and form BHs without unbinding the binary, leaving a GW source behind. BH in binaries can also form separately, from two uncorrelated stars, and later meet because of dynamical interactions (right). The latter class of models typically requires a dense stellar environment to efficiently assemble GW sources.

There are two main classes of stellar-based BH-binary formation models, depending on whether the two BHs have spent all their lives together as a binary star [9], or they rather evolved separately and later met to form a BH binary [10] (see Fig. 1 for a schematic diagram). In the first scenario (Fig. 1, left panel), two massive stars forming a binary system *in the field* both evolve into BHs and leave a compact binary behind. There are several key steps involved in this class of models, notably a common envelope phase and natal kicks imparted to the BHs at birth. In the second class of models (Fig. 1, right panel), a dense stellar environment such as a globular or open cluster is invoked to assemble BHs through dynamical interactions.

We face a pressing question in GW astrophysics: can we distinguish between these formation channels using current and future BH binary observations? Can we infer some of the key parameters entering these models? In other terms: *do BH binaries remember how they form?*

The BH mass distributions predicted by several formation models tend to overlap, meaning that $O(100)$ observations will be necessary before convincing constraints can be extracted using mass measurements only [11, 12]. Eccentricity and redshift measurements could provide strong constraints for specific formation models (cf. [13] and [14], respectively), but are perhaps not going to be very informative for the most favorable scenarios. Spins, on the other hand, are arguably the cleanest observables to distinguish BH formation channels. If the two BHs were together as a binary star, the relative orientations of their spins and the binary's orbital angular momentum are expected to reflect processes able to (mis)align the stellar spins (such as BH natal kicks and tidal interactions [15, 16, 17]). Conversely, if BHs form in clusters, their spin orientation is expected to be uncorrelated and isotropically distributed.

Here I present some first attempts to exploit the spin dynamics to extract astrophysical information from current and future BH binary observations. Sec. 2 summarizes the phenomenology of two-spin BH binaries and introduces the spin morphology as a tool to directly infer the spin orientations at BH formation from those observed by LIGO/Virgo at ~ 100 Hz. Sec. 3 briefly describes how spins can be used to extract information on BH natal kicks and the occurrence of multiple merger generations. We draw our conclusions in Sec. 4. Unless otherwise stated, we use geometric units $c = G = 1$.

2. Timescales and morphologies

The phenomenology of spinning BH binaries is better understood by separating the various timescales of the problem. The two BHs orbit about each other on a timescale $t_{\rm orb} \propto (r/M)^{3/2}$, the spins and the orbital angular momentum precess on a timescale $t_{\rm pre} \propto (r/M)^{5/2}$, while the orbit shrinks on $t_{\rm rad} \propto (r/M)^4$ [18] (here r is the binary separation and $M = m_1 + m_2$ is the total mass). In the post-Newtonian (PN) regime $r/M \gg 1$ and therefore $t_{\rm orb} \ll t_{\rm pre} \ll t_{\rm rad}$. The first inequality $t_{\rm orb} \ll t_{\rm pre}$ has been used since the very early developments of the PN formalism, leading to the orbit-averaged formulation of the PN equations of motion. In general, the relative orientations of the spins $\mathbf{S_i} = \chi_i m_i^2 \hat{\mathbf{S}}_\mathbf{i}$ (with $i = 1, 2$) and the binary's angular momentum $\mathbf{L} = m_1 m_2 \sqrt{r/M}\hat{\mathbf{L}}$ depend on three variables, which can be chosen the be the angles θ_i between $\mathbf{S_i}$ and $\mathbf{L}$, and the angle $\Delta\Phi$ between $\hat{\mathbf{S}}_\mathbf{1} \times \hat{\mathbf{L}}$ and $\hat{\mathbf{S}}_\mathbf{2} \times \hat{\mathbf{L}}$ (c.f. Fig. 1 in [19]). Only recently [19, 20], we presented an analysis of the BH binary dynamics which fully exploits the second inequality $t_{\rm pre} \ll t_{\rm rad}$, and allows us to formulate the equations in a *precession-averaged* fashion. This was made possible by the presence of two additional constants of motion: (i) the effective spin $\chi_{\rm eff}$ (or ξ in the notation of [19, 20]) is constant on both $t_{\rm pre}$ and $t_{\rm rad}$ [21], and (ii) the magnitude $J = |\mathbf{L} + \mathbf{S_1} + \mathbf{S_2}|$ is constant on $t_{\rm pre}$ and slowly evolves on $t_{\rm rad}$. On the precession time, the entire dynamics can be described in terms of the single variable $S = |\mathbf{S_1} + \mathbf{S_2}|$.

The equations of motion can now be solved with straightforward numerical integrations [19], or even analytically [22]. The precession dynamics can be classified in terms of the qualitative evolution of the spin angles θ_1, θ_2 and $\Delta\Phi$. While θ_1 and θ_2 always evolve monotonically during each precession cycle, three different *morphologies* can be found in $\Delta\Phi$. Orbits of $\Delta\Phi$ can circulate in the whole range $[0, \pi]$, librate about 0 (never reaching π) or librate about π (never reaching 0). The crucial point here is that, while the spin angles θ_1, θ_2 and $\Delta\Phi$ vary on the short timescale $t_{\rm pre}$, the spin morphology does not. It is only set by the values of χ_1, χ_2, $q = m_2/m_1$, $\chi_{\rm eff}$ and J, and is therefore constant on $t_{\rm pre}$. Secular variations of J on $t_{\rm rad}$ might cause transitions between the three morphologies, generically acting towards bringing librating binaries towards the two circulating morphologies.

Quantities subject to spin precession are typically quoted at some reference frequency in GW searches (set to $f_{\rm ref} = 20$ Hz in current LIGO analyses). We argue one should always try to estimate quantities varying on the longest timescales of the problem, so that the values reported better characterize the observed systems. From this point of view, the effective spin parameter $\chi_{\rm eff}$ is an excellent choice, because it is constant on both $t_{\rm pre}$ and $t_{\rm rad}$ to high PN order. The spin morphology is also an interesting quantity, because it only varies on $t_{\rm rad}$. On the contrary, the spin angles vary on the shorter timescale $t_{\rm pre}$.

The spin morphology, therefore, encodes information on the precessional dynamics of the binary, but does not vary on the precessional time. Moreover, if the spin morphology is estimated at, say, 20 Hz, (well within the detector sensitivity window) it can easily be used to infer information on the spin orientations at BH formation and thus constrain astrophysical models. This point is illustrated in Fig. 2. We evolved a sample of 500 BH binaries with $\chi_1 = \chi_2 = 1$ and $q = 0.8$ from $r = 10^6 M$ to $r = 10M$. Each binary is color-coded according to its spin morphology measured at $r = 10M$. The spin directions θ_i at small separations carry little information on the initial configuration of the binary. Conversely, the spin morphology directly tracks the spin orientation at early times: binaries of different colors cluster in different and well-defined regions of the (θ_1, θ_2) plane. By measuring the spin morphology in the LIGO band, we will be able to robustly reconstruct information on the spin configuration when the BHs formed.

This idea can be used to constrain specific formation mechanisms that preferentially populate some regions of the (θ_1, θ_2) parameter space. An early exploration was presented in [17], where we showed that stellar-mass BH binaries formed in the field tend to belong to different morphologies if they (do not) undergo mass transfer and tidal interactions during their lives as massive stars.

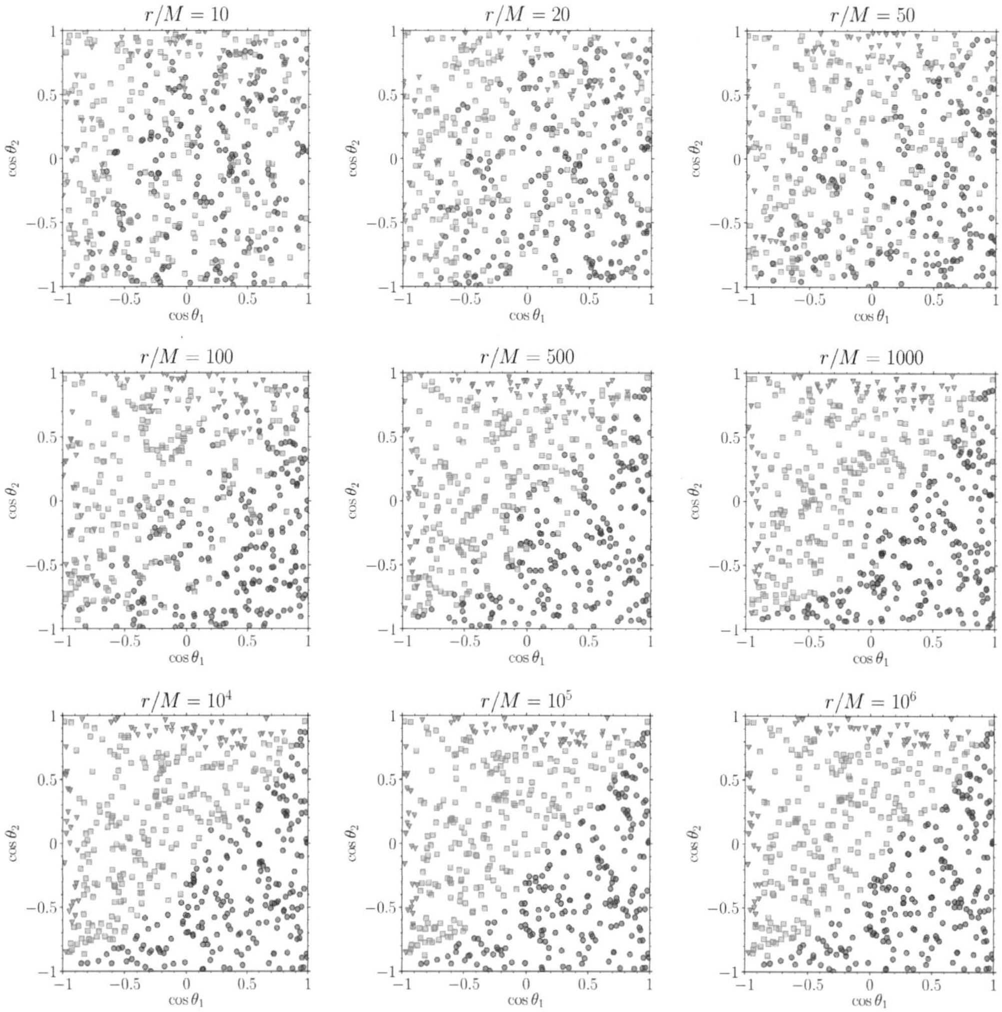

Figure 2. Precession-averaged PN evolutions of BH binaries. We evolve 500 systems with $\chi_1 = \chi_2 = 1$ and $q = 0.8$ with isotropic spin directions at large separations. The evolution is shown backwards, from $r = 10M$ (top-left panel) to $r = 10^6 M$ (bottom-right panel), in the plane defined by the two spin orientations $\cos\theta_i = \hat{\mathbf{S}}_\mathbf{i} \cdot \hat{\mathbf{L}}$. Each binary is colored according to its spin morphology at $r = 10M$: blue for binaries librating about $\Delta\Phi = 0$, green for binaries circulating in $\Delta\Phi \in [0, \pi]$ and red for binaries librating about $\Delta\Phi = \pi$. Binaries with a given morphology at small separation (detection) originate from precise regions in the (θ_1,θ_2) plane at large separation (formation). This figure was produced using the public python module PRECESSION [23]. An animated version is available at www.davidegerosa.com/spinprecession.

3. Spin constraints on black-hole natal kicks and multiple generations

Although weakly informative, current LIGO measurements of BH spins already provide some interesting indications. In particular, GW150914, GW170104, GW170814 and LVT151012 are all consistent with zero spins [1, 3, 4]. The opposite is true for GW151226, where at least one of the two BHs must have been spinning before merger [2, 24] . These few measurements can be used to place (model-dependent) constraints on the astrophysics of BH binaries. Here we summarize two of our recent studies on the topic.

- The "boxing day" event, GW151226, can be used to infer the magnitude of the recoil velocities that BHs are expected to receive at birth [25]. At core collapse and BH formation, asymmetric mass and neutrino emission might impart a recoil to the newly-formed compact object. While natal kicks imparted to neutron stars are well constrained by pulsar observations [26], their occurrence on BHs is still uncertain [27, 28, 29]. The primary BH of GW151226 likely had non-zero spin, modestly misaligned with the angular momentum ($25^\circ \lesssim \theta_1 \lesssim 80^\circ$) [2, 25]. In a field formation scenario where stellar spins are initially aligned to the orbital angular momentum, this spin misalignment directly tracks the orbital-plane tilt caused by the first recoil [15, 17, 25]. We found that a kick $v_k \sim v$ (where v is the orbital velocity at BH formation) must be imparted in order to produce a misalignment of $25^\circ \lesssim \theta_1 \lesssim 80^\circ$, consistent with that observed for GW151226. This constraint can be converted into physical units by averaging over a stellar population [30, 31]. We find only values $v_k \gtrsim 45$ km/s can reproduce the observed BH misalignment in $> 5\%$ of the realizations. Our analysis was later refined in [32], where more elaborate, but still model-dependent, constraints are presented using masses *and* spin information of all the events available at the time.
- One possibility is that the BHs we have been observing do not come directly from the collapse of stars ("first generation"), but rather result from previous BH mergers ("second generation") [33]. While it is certainly true that some astrophysical interactions are expected to be at play to assemble multiple merger generations (e.g. [34, 35]), our approach is agnostic on the specific formation channel. This direction is therefore orthogonal, but complementary, to the usual field vs. cluster debate (Fig. 1). Once more, the secret to discriminate these models lies in the BH spins. Spins of BHs resulting from a previous merger are mainly set by the angular momentum of the progenitor binary at plunge, with the component spins playing a subdominant role. In practice, this means that the spin distribution of second-generation BHs is expected to be highly peaked at $\chi \sim 0.7$ [36, 33]. Multiple merger generations are expected to not only affect the spin distributions, but also BH masses and merger redshifts. On average, second-generation BHs should be more massive, and merge later. In [33], we combined these insights with Bayesian model-selection techniques to constrain the occurrence of multiple mergers with current and future GW data. We showed that BH binaries from LIGO's first observing run already provide a 1 to 2σ constraint favoring first generation mergers. With $20 - 100$ observations we will be able to constrain these models at 5σ in $> 90\%$ of the realizations.

4. Conclusions

BH spins are precious observables to infer the astrophysics of BH binaries from GW observations. We showed how information on the BH spins at formation is not erased by the large number of precession cycles that sources undergo before entering the LIGO/Virgo band. Some memory is preserved and encoded into the spin morphologies. The first GW detections are already providing us interesting constraints on the astrophysics of massive stars and BH evolution. We described two examples, where we pioneered the use of GW data to constrain specific formation and evolution mechanisms: (i) the observed misalignment of GW151226 can be used to obtain a (model dependent) constraint on BH natal kicks; and (ii) mass, redshift, and spin distributions can be used to constrain scenarios where multiple merger generations are responsible for the observed events.

This is just the beginning of a new era in astronomy and relativity. The future is bright, with more events, better detectors, new kind sources, more electromagnetic coincidences and serendipitous discoveries all lying just beyond the horizon.

Acknowledgments

These findings were presented in a plenary talk at the 12th Edoardo Amaldi Conference on Gravitational Waves, following the awarding ceremony of the 2016 Stefano Braccini Thesis Prize. D.G. would like to thank the Gravitational Wave International Committee and all friends of Stefano Braccini for this award, and encourage future graduates to submit their Theses for consideration. D.G thanks E. Berti, M. Kesden, R. O'Shaughnessy, U. Sperhake, and D. Wysocki, who all contributed to the research presented in these proceedings. D.G. is supported by NASA through Einstein Postdoctoral Fellowship Grant No. PF6-170152 awarded by the Chandra X-ray Center, which is operated by the Smithsonian Astrophysical Observatory for NASA under Contract NAS8-03060.

References

[1] Abbott, B P *et al* (LIGO and Virgo Scientific Collaboration) 2016 *PRL* **116** 061102 (arXiv:1602.03837)
[2] Abbott, B P *et al* (LIGO and Virgo Scientific Collaboration) 2016 *PRL* **116** 241103 (arXiv:1606.04855)
[3] Abbott, B P *et al* (LIGO and Virgo Scientific Collaboration) 2017 *PRL* **118** 221101 (arXiv:1706.01812)
[4] Abbott, B P *et al* (LIGO and Virgo Scientific Collaboration) 2017 *PRL* **119** 141101 (arXiv:1709.09660)
[5] Abbott, B P *et al* (LIGO and Virgo Scientific Collaboration) 2016 *PRX* **6** 041015 (arXiv:1606.04856)
[6] Özel F, Psaltis D, Narayan R and McClintock J E 2010 *ApJ* **725** 1918–1927 (arXiv:1006.2834)
[7] Abbott, B P *et al* (LIGO and Virgo Scientific Collaboration) 2016 *ApJ* **818** L22 (arXiv:1602.03846)
[8] Lamberts A, Garrison-Kimmel S, Clausen D R and Hopkins P F 2016 *MNRAS* **463** L31–L35 (arXiv:1605.08783)
[9] Postnov K A and Yungelson L R 2014 *Living Reviews in Relativity* **17** 3 (arXiv:1403.4754)
[10] Benacquista M J and Downing J M B 2013 *Living Reviews in Relativity* **16** 4 (arXiv:1110.4423)
[11] Stevenson S, Ohme F and Fairhurst S 2015 *ApJ* **810** 58 (arXiv:1504.07802)
[12] Zevin M, Pankow C, Rodriguez C L, Sampson L, Chase E, Kalogera V and Rasio F A 2017 *ApJ* **846** 82 (arXiv:1704.07379)
[13] Antonini F and Perets H B 2012 *ApJ* **757** 27 (arXiv:1203.2938)
[14] Belczynski K, Ryu T, Perna R, Berti E, Tanaka T L and Bulik T 2017 *MNRAS* **471** 4702–4721 (arXiv:1612.01524)
[15] Kalogera V 2000 *ApJ* **541** 319–328 (arXiv:astro-ph/9911417)
[16] Hut P 1981 *A&A* **99** 126–140
[17] Gerosa D, Kesden M, Berti E, O'Shaughnessy R and Sperhake U 2013 *PRD* **87** 104028 (arXiv:1302.4442)
[18] Apostolatos T A, Cutler C, Sussman G J and Thorne K S 1994 *PRD* **49** 6274–6297
[19] Gerosa D, Kesden M, Sperhake U, Berti E and O'Shaughnessy R 2015 *PRD* **92** 064016 (arXiv:1506.03492)
[20] Kesden M, Gerosa D, O'Shaughnessy R, Berti E and Sperhake U 2015 *PRL* **114** 081103 (arXiv:1411.0674)
[21] Racine É 2008 *PRD* **78** 044021 (arXiv:0803.1820)
[22] Chatziioannou K, Klein A, Cornish N and Yunes N 2017 *PRL* **118** 051101 (arXiv:1606.03117)
[23] Gerosa D and Kesden M 2016 *PRD* **93** 124066 (arXiv:1605.01067)
[24] Vitale S, Gerosa D, Haster C J, Chatziioannou K and Zimmerman A 2017 (arXiv:1707.04637)
[25] O'Shaughnessy R, Gerosa D and Wysocki D 2017 *PRL* **119** 011101 (arXiv:1704.03879)
[26] Hobbs G, Lorimer D R, Lyne A G and Kramer M 2005 *MNRAS* **360** 974–992 (arXiv:astro-ph/0504584)
[27] Repetto S, Davies M B and Sigurdsson S 2012 *MNRAS* **425** 2799–2809 (arXiv:1203.3077)
[28] Martin R G, Tout C A and Pringle J E 2010 *MNRAS* **401** 1514–1520 (arXiv:0910.0018)
[29] Janka H T 2012 *Annual Review of Nuclear and Particle Science* **62** 407–451 (arXiv:1206.2503)
[30] Sana H, de Mink S E, de Koter A, Langer N, Evans C J, Gieles M, Gosset E, Izzard R G, Le Bouquin J B and Schneider F R N 2012 *Science* **337** 444 (arXiv:1207.6397)
[31] Belczynski K, Holz D E, Bulik T and O'Shaughnessy R 2016 *Nature* **534** 512–515 (arXiv:1602.04531)
[32] Wysocki D, Gerosa D, O'Shaughnessy R, Belczynski K, Gladysz W, Berti E, Kesden M and Holz D 2017 (arXiv:1709.01943)
[33] Gerosa D and Berti E 2017 *PRD* **95** 124046 (arXiv:1703.06223)
[34] McKernan B, Ford K E S, Bellovary J, Leigh N W C, Haiman Z, Kocsis B, Lyra W, MacLow M M, Metzger B, O'Dowd M, Endlich S and Rosen D J 2017 (arXiv:1702.07818)
[35] Samsing J and Ilan T 2017 (arXiv:1709.01660)
[36] Berti E and Volonteri M 2008 *ApJ* **684** 822-828 (arXiv:0802.0025)